Praise for *The World in a Grain*

"[An] impassioned and alarming report on sand. . . . In Beiser's artful telling, the planet is caught up in a vicious, sand-fueled cycle."
—*The Washington Post*

"Beiser peppers research with first-person interviews in an engaging and nuanced introduction to the ways sand has shaped the world . . . stunning."
—NPR

"Beiser's eye-opening study clarifies the science and the huge role of sand in heavy and high-tech industry. Perhaps most compelling is his exposé of sand mining, which obliterates islands, destroys coral reefs and marine biodiversity, and threatens livelihoods. A powerful lens on an underreported environmental crisis."
—*Nature*

"Whether in Chippewa Falls, Wisconsin, or India, [Beiser] exhibits a flare for detailing the human drama through prose."
—*Los Angeles Review of Books*

"I thought I knew the basics of sustainability, but this lucid, eye-opening book made me feel like a dolt in the best possible aha-moment way: I'd simply never registered how much of the contemporary world—our concrete and glass buildings and asphalt roads and silicone-based digital devices and so much more—is entirely, voraciously sand-dependent. And the looming global sand *crisis*: Who knew?"
—Kurt Andersen, author of *Fantasyland: How America Went Haywire: A 500-Year History*

"*The World in a Grain* is nothing less than one of the best reporters working today unpacking the literal foundations of civilization. Everything we are, everywhere we live, is built on or out of sand, and Vince Beiser tells the best story of where that sand comes from, who moves it, and what they build from it. It's a whole new way of seeing the world."
—Adam Rogers, author of *Proof: The Science of Booze*

"A riveting, wonderfully written investigation into the many kinds of castles the world has built out of sand. You'll find something new, and something fascinating, on every page. Perhaps even in every paragraph."
—Nicholas Thompson, author of *The Hawk and the Dove*

"A fresh history of 'the most important solid substance on Earth, the literal foundation of modern civilization.' Books on a single, familiar topic (salt, cod, etc.) have an eager audience, and readers will find this an entirely satisfying addition to the genre." —*Kirkus Reviews*

"The book is at its urgent best in chapters on the black market in sand and the sand mafias that brutally exercise control over resources. . . . Breezily written and with insights on every page, this is an eye-opening look at a resource too often taken for granted." —*Publishers Weekly*

"A rich study of one of the world's most abundant natural resources: sand. With a balance of statistics, science, history, on-the-scene reporting, and some healthy environmental skepticism, *The World in a Grain* highlights the ways this ubiquitous global commodity has been essential to human development and advancement." —*Shelf Awareness*

"Modern life, as Vince Beiser compellingly explains, is literally made of sand. Yet we have been so profligate with this seemingly inexhaustible resource that for many uses in many parts of the world we are running out. *The World in a Grain* is a chronicle of innovation and greed and heedless waste—in brief, the story of civilization."
—David Owen, author of *Where the Water Goes*

"Sand shortage? Black market in sand? Secret sand heists? Who knew? I certainly didn't before reading this lively and eye-opening book about a material I'd always assumed almost infinite. Vince Beiser shows, with great skill, that this key component of our fragile, overconsuming planet is something we need to better understand, conserve, and protect."
—Adam Hochschild, author of *King Leopold's Ghost* and *Bury the Chains*

The World in a Grain

The Story of Sand and
How It Transformed
Civilization

VINCE BEISER

RIVERHEAD BOOKS
New York

RIVERHEAD BOOKS
An imprint of Penguin Random House LLC
penguinrandomhouse.com

Parts of several chapters of this book first appeared, in different form,
in *Wired, The New York Times, The Guardian, Pacific Standard,* NationalGeographic.com,
and *Mother Jones.*

The Library of Congress has catalogued the Riverhead hardcover edition as follows:

Names: Beiser, Vince, author.
Title: The world in a grain : the story of sand and how it transformed
civilization / Vince Beiser.
Description: New York : Riverhead Books, 2018. | Includes bibliographical
references and index.
Identifiers: LCCN 2017053122| ISBN 9780399576423 (hardcover) |
ISBN 9780399576430 (ebook)
Subjects: LCSH: Sand. | Technology and civilization.
Classification: LCC TA455.S3 B45 2018 | DDC 620.1/91—dc23
LC record available at https://lccn.loc.gov/2017053122
p.   cm.

First Riverhead hardcover edition: August 2018
First Riverhead trade paperback edition: August 2019
Riverhead trade paperback ISBN: 9780399576447

Printed in the United States of America

Book design by Gretchen Achilles

For Kaile, Adara, and Isaiah.
I love you more than there are grains of sand
in the whole wide world.

CONTENTS

CHAPTER 1

The Most Important Solid Substance on Earth

This book is about something most of us barely ever think about and yet can't live without. It is about the most important solid substance on Earth, the literal foundation of modern civilization.

It is about sand.

Sand? Why is this humblest of materials, something that seems as trivial as it is ubiquitous, so significant?

Because sand is the main material that modern cities are made of. It is to cities what flour is to bread, what cells are to our bodies: the invisible but fundamental ingredient that makes up the bulk of the built environment in which most of us live.

Sand is at the core of our daily lives. Look around you right now. Is there a floor beneath you, walls around, a roof overhead? Chances are excellent they are made at least partly out of concrete. And what is concrete? It's essentially just sand and gravel glued together with cement.

Take a glance out the window. All those other buildings you see are also made from sand. So is the glass in that window. So are the miles of asphalt roads that connect all those buildings. So are the silicon chips that are the brains of your laptop and smartphone. If you're in downtown San Francisco, in lakefront Chicago, or at Hong Kong's international airport, the very ground beneath you is likely artificial, manufactured with sand dredged up from underwater. We humans bind together countless trillions of grains of sand to build towering structures, and we break apart the molecules of individual grains to make tiny computer chips.

Some of America's greatest fortunes were built on sand. Henry J. Kaiser, one of the wealthiest and most powerful industrialists of twentieth-century America, got his start selling sand and gravel to road builders in the Pacific Northwest. Henry Crown, a billionaire who once owned the Empire State Building, began his own empire with sand dredged from Lake Michigan that he sold to developers building Chicago's skyscrapers. Today the construction industry worldwide consumes some $130 billion[1] worth of sand each year.

Sand lies deep in our cultural consciousness. It suffuses our language. We draw lines in it, build castles in it, hide our heads in it. In medieval Europe (and a classic Metallica song), the Sandman helped ease us into sleep. In our modern mythologies, the Sandman is a DC superhero and a Marvel supervillain. In the creation myths of indigenous cultures from West Africa to North America, sand is portrayed as the element that gives birth to the land.[2] Buddhist monks and Navajo artisans have painted with it for centuries. "Like sands through the hourglass, so are the days of our lives," intone the opening credits of a classic American soap opera. William Blake encouraged us to "see a world in a grain of sand." Percy Bysshe Shelley reminded us that even the mightiest of kings end up dead and forgotten, while around them only "the lone and level sands stretch far away." Sand is both minuscule and infinite, a means of measurement and a substance beyond measuring.

Sand has been important to us for centuries, even millennia.

People have used it for construction since at least the time of the ancient Egyptians. In the fifteenth century, an Italian artisan figured out how to turn sand into fully transparent glass, which made possible the microscopes, telescopes, and other technologies that helped drive the Renaissance's scientific revolution.

But it was only with the advent of the modern industrialized world, in the decades just before and after the turn of the twentieth century, that people really began to harness the full potential of sand and begin making use of it on a colossal scale. It was during this period that sand went from being a resource used for widespread but artisanal purposes to becoming the essential building block of civilization, the key material used to create mass-manufactured structures and products demanded by a fast-growing population.

At the dawn of the twentieth century, almost all of the world's large structures—apartment blocks, office buildings, churches, palaces, fortresses—were made with stone, brick, clay, or wood. The tallest buildings on Earth stood fewer than ten stories high. Roads were mostly paved with broken stone, or more likely, not paved at all. Glass in the form of windows or tableware was a relatively rare and expensive luxury. The mass manufacture and deployment of concrete and glass changed all that, reshaping how and where people lived in the industrialized world.

Then in the years leading up to the twenty-first century, the use of sand expanded tremendously again, to fill needs both old and unprecedented. Concrete and glass began rapidly expanding their dominion from wealthy Western nations to the entire world. At roughly the same time, digital technology, powered by silicon chips and other sophisticated hardware made with sand, began reshaping the global economy in ways gargantuan and quotidian.

Today, your life depends on sand. You may not realize it, but sand is there, making the way you live possible, in almost every minute of your day. We live in it, travel on it, communicate with it, surround ourselves with it.

Wherever you woke up this morning, chances are good it was in a building made at least partly out of sand. Even if the walls are made of brick or wood, the foundation is most likely concrete. Maybe it's also plastered with stucco, which is mostly sand. The paint on your walls likely contains finely ground silica sand to make it more durable, and may include other forms of high-purity sands to increase its brightness, oil absorption, and color consistency.[3]

You flicked on the light, provided by a glass bulb made from melted sand. You meandered to the bathroom, where you brushed your teeth over a sink made of sand-based porcelain, using water filtered through sand at your local purification plant. Your toothpaste likely contained hydrated silica,[4] a form of sand that acts as a mild abrasive to help remove plaque and stains.

Your underwear snapped into place thanks to an elastic made with silicone, a synthetic compound also derived from sand. (Silicone also helps shampoo make your hair shinier, makes shirts less wrinkle-prone, and reinforced the boot sole with which Neil Armstrong made the first footprint on the moon. And yes, most famously, it has been used to enhance women's busts for more than fifty years.)

Dressed and ready, you drove to work on roads made of concrete or asphalt. At the office, the screen of your computer, the chips that run it, and the fiber-optic cables that connect it to the Internet are all made from sand. The paper you print your memos on is probably coated with a sand-based film that helps it absorb printer ink. Even the glue that makes your sticky notes stick is derived from sand.

At day's end, you flopped down with a glass of wine. Guess

what? Sand was used to make the bottle, the glass, and even the wine. Wine is sometimes made with a dash of colloidal silica, a gel form of silicon dioxide used as a "fining" agent to improve the beverage's clarity, color stability, and shelf life.

Sand, in short, is the essential ingredient that makes modern life possible. Without sand, we couldn't have contemporary civilization.

And believe it or not, we are starting to run out.

Though the supply might seem endless, usable sand is a finite resource like any other. (Desert sand generally doesn't work for construction; shaped by wind rather than water, desert grains are too round to bind together well.)[5] We use more of this natural resource than of any other except air and water. Humans are estimated to consume nearly 50 billion tons of sand and gravel every year.[6] That's enough to blanket the entire state of California. It's also twice as much as we were using just a decade ago.

Today, there is so much demand for sand that riverbeds and beaches around the world are being stripped bare of their precious grains. Farmlands and forests are being torn up. And people are being imprisoned, tortured, and murdered. All over sand.

The key factor driving our world's unprecedented consumption of this humblest of materials is this: the number and size of cities is exploding. Every year there are more and more people on the planet, and every year more and more of them move to cities, especially in the developing world.

The scale of this migration is staggering. In 1950, some 746 million people—less than one-third of the world's population—lived in cities. Today, the number is 4.2 billion, more than half of all the people on Earth.[7] The United Nations predicts that another 2.5 billion will join them in the next three decades.[8] The global urban population is rising by about 65 million people annually; that's the equivalent of adding eight New York Citys to the planet every single year.

To build these cities of concrete, asphalt, and glass, humans are pulling sand out of the ground in exponentially increasing amounts. The overwhelming bulk of it goes to make concrete, by far the world's most important building material. In a typical year, according to the United Nations Environment Programme, the world uses enough concrete to build a wall 88 feet high and 88 feet wide right around the equator.[9] China alone used more cement between 2011 and 2013 than the United States used in the entire twentieth century.[10]

There is such intense need for certain types of construction sand that places like Dubai, which sits on the edge of an enormous desert in the Arabian Peninsula, are importing sand from Australia. That's right: exporters in Australia are literally selling[11] sand to Arabs.[12]

What is sand, anyway? That simple syllable comprises a panoply of tiny objects of many shapes and sizes made of many different substances. As defined by the Udden-Wentworth scale, the most commonly used geologic standard, the term *sand* encompasses loose grains of any hard material with a diameter between 2 and 0.0625 millimeters. That means the average grain of sand is a tad larger than the width of a human hair. Those grains can be made by glaciers grinding up stones, by oceans degrading seashells and corals (many Caribbean beaches are made of decomposed shells),[13] even by volcanic lava chilling and shattering upon contact with air or water. (That's where Hawaii's black sand beaches come from.)[14]

Nearly 70 percent of all sand grains on Earth, however, are

quartz. These are the ones that matter most to us. Quartz is a form of silicon dioxide, or SiO_2, also known as silica. Its components, silicon and oxygen, are the most abundant elements in the Earth's crust, so it's no surprise that quartz is one of the most common minerals on Earth.[15] It is found abundantly in the granite and other rocks that form the world's mountains and other geologic features.

Most of the quartz grains we use were formed by erosion. Wind, rain, freeze-thaw cycles, microorganisms, and other forces eat away at mountains and other rock formations, breaking grains off their exposed surfaces. Rain then washes those grains downhill, sweeping them into rivers that carry countless tons of them far and wide. This waterborne sand accumulates in riverbeds, on riverbanks, and on the beaches where the rivers meet the sea. Over the centuries, rivers periodically overflow their banks and shift their courses, leaving behind huge deposits of sand in what has become dry land.[16] Quartz is tremendously hard, which is why quartz grains survive this long, bruising journey intact while other mineral grains disintegrate.

Over millions of years, sands are often buried under newer layers of sediment, uplifted into new mountains, then eroded and transported once again. "Sand grains have no souls, but they are reincarnated," writes geologist Raymond Siever[17] in his book *Sand*. "Each cycle of deposition, burial, uplift and erosion renews the sand grains and rounds each grain a little more." The average time for this cycle is 200 million years. The next time you dump sand out of your shoes, give those grains a little respect: they may predate the dinosaurs.

In the wild, quartz always comes mixed with bits of other materials: iron, feldspar, whatever other minerals prevail in the local geology. (Pure quartz is transparent, but quartz grains are often stained by oxidation. That coloring, plus the presence of other types of grains, is why most beaches and sand deposits you see are various shades of yellow or brown.) A certain amount of those other substances need to be filtered out before the sand can be used to make concrete, glass, or other products.

You can think of sand sort of like a colossal army, or a group of related armies, made up of quintillions of tiny soldiers. Only these armies are deployed not to kill, but to create. Rather than destroy, these soldiers build structures and products and perform services for us.

At first glance, sand grains, like uniformed troops, all look pretty much the same. In fact, though, there are many different types, with different attributes, strengths, and weaknesses, which in turn determine the uses to which they can be put. Some are prized for their hardness, some for their pliancy; some for their roundness, some for their angularity; some for their color, some for their purity. Some sands, like specially chosen commandos, are put through elaborate physical or chemical processes to alter their capabilities, or are combined with other materials to perform tasks they could not in their original state.

Construction sand—the hard, angular grains used primarily to make concrete—are the infantry of this army. This kind of sand is abundant, easily found, and not especially pure. Its grains are mainly quartz, but include other minerals, which vary depending on where the sand was mined. Construction sand can be found in virtually every country, often mixed with its indispensable partner, gravel. The construction industry refers to sand and gravel together as *aggregate*; the difference between sand and gravel is mainly just size. These particles are drafted into service from riverbeds, beaches, or land quarries. Sand and gravel aggregates are put to work together to make concrete, while sand is deployed on its own to make other construction materials like mortar, plaster, and roofing components.

Marine sands—the naval wing of the army, found on the ocean floor—are of similar composition, making them useful for artificial land building, such as Dubai's famous palm-tree-shaped manmade islands. These underwater grains can also be used for concrete, but that requires washing the salt off them—an expensive step most contractors would rather avoid.

Silica sands are purer—at least 95 percent[18] silica—and are found in fewer places than construction or marine sand. Also called industrial sands, they're the Special Forces of the sand army, capable of being put to more sophisticated purposes than the average foot soldier. These are the sands you need to make glass. Higher-purity sands are especially prized: the sands of north-central France's Fontainebleau region, for instance, are upward of 98 percent pure silica. Europe's finest glassmakers have relied on them for centuries. Silica sands are also used to help make molds for metal foundries, add luster to paint, and filter the water in swimming pools,[19] among many other tasks. Some of the unique properties of industrial sands suit them for highly specific jobs. The silica sands of western Wisconsin, for instance, have a particular shape and structure that make them ideal for use in fracking for oil and gas.

Then there is the SEAL Team Six of the silica world: relatively small amounts of extremely high-purity quartz, a tiny, elite group possessed of rare attributes that enable them to perform extraordinary feats. These particles are made into high-tech equipment essential for manufacturing computer chips. Some are also used to create the sparkling sand traps of exclusive golf courses or to line Persian Gulf horse-racing tracks—like elite commandos taking jobs as a rich man's bodyguard.

For the most part, we don't draft desert sands into our service. The grains found in deserts are mostly too round to use for construction. The reason is that wind is harsher than water. In a river, water cushions the impact of the grains tumbling against one another. In a desert, they just bang full force into one another, rounding off their corners and angles.[20] Round objects don't lock together as nicely as angular ones. It's like the difference between trying to pile up a bunch of marbles as opposed to stacking up a bunch of blocks.

We summon up these tiny soldiers in many different ways and in many places. In some places, multinational companies dredge sand from riverbeds or gouge it out of hillsides with massive machines. In others, local people haul it away with shovels and pickup trucks.

Generally speaking, sand mining is a relatively low-tech industry. The basic machinery involved hasn't changed much since the 1920s. Sand from the beds of rivers and lakes is dredged up with suction pumps, or clamshell claws mounted on floating platforms, or ships equipped with scoops set on conveyor belts. Underwater sands are easier to mine, since there's no intervening earth, known as *overburden*, to scrape away. They also come largely cleansed of dust-sized particles. On land, sand is usually quarried from open pits. Sometimes that requires using explosives and crushing machines to break apart sandstone, rock made of sand that has been glued together over the millennia by naturally occurring cements. Regardless of its source, the raw sand needs to be washed and run through a series of screens to sort it by size.

Because sand is so common, there are sand mines all over the place, in almost every country. There is no one key source, no Saudi Arabia of sand. Much of the extraction of sand is carried out by relatively small regional companies. In the United States, some

4,100 companies and government agencies harvest aggregate from about 6,300 locations in all fifty states.[21] The breakdown is similar in Western Europe.[22]

Though it's often carried out on a small, seemingly insignificant scale, there's no escaping the fact that sand mining is *mining*; it's an extractive industry that inevitably affects the natural world. All those thousands of small mines, together with many larger ones, add up to a colossal impact. Sand mining tears up wildlife habitat, fouls rivers, and destroys farmland. The damage can be mitigated. Some companies are more conscientious than others, some extraction methods are more disruptive than others, and some governments are more vigilant than others. But everywhere, the process of pulling sand from the earth causes at best a little damage, and at worst, catastrophe.

Perhaps the only place where most people really appreciate sand—or even think about it—is the beach. Those beloved strips of sun-kissed shore are on the front lines of the global battle for sand, and they are taking heavy fire.

The beach near the tiny town of Marina, California, a couple of hours south of San Francisco, is a broad stretch of wild, undeveloped sand sloping into the foaming waves of the Pacific Ocean. Much of its miles-long expanse is designated as a state park. Hidden away behind high dunes bedecked with green and orange succulents, it's a postcard-perfect slice of natural beauty. And it is gradually disappearing.

"This is the fastest-eroding shoreline in California," said Ed Thornton, a retired coastal engineer and former professor with the Naval Postgraduate School in nearby Monterey, to a crowd of protesters gathered on the beach in early 2017. "We're losing eight acres a year of pristine shore, some of the most beautiful in the world. It's because of sand mining."

The demonstration was being held near a hulking dredge operated by Cemex, a global construction firm based in Mexico. At the time, this machine was sucking up an estimated 270,000 cubic meters of sand from a tide-filled lagoon every year. The grains were to be bagged and sold to contractors across the country for sand blasting[23] and lining oil and gas wells.

For most of the twentieth century there were many such ocean sand mines along the California coast. But in the late 1980s the federal government shut them down because it had become clear the loss of sand was severely eroding the Golden State's famous beaches. the Cemex operation, however, kept running thanks to a legal loophole: the dredging area appeared to sit above the mean high tide line, putting it out of federal jurisdiction. Activists and local legislators fought for years to shut the mine down. A few months after that demonstration on the beach, they finally won: Cemex agreed to phase out the dredging by late 2020.

That still leaves at least one sand mine in operation that may be damaging California's coastal areas, however. Environmentalists are battling in court to stop sand dredging in San Francisco Bay, saying it is causing the erosion of a nearby ocean beach and endangering bird habitat.[24]

In other parts of the world, the impact of sand miners on beaches is more clear-cut. They're actively stealing them. Thieves in Jamaica made off with 1,300 feet of white sand from one of the island's finest beaches in 2008. Smaller-scale beach-sand looting is ongoing in Morocco, Algeria, Russia, Italy, and many other places around the world. In Florida, southern France, and many other vacation hot spots, beaches are shrinking thanks to other forms of human interference, as we'll see in chapter 7.

The damage being done to beaches is only one facet, and not even the most dangerous one, of the damage being done by sand mining around the world.

Sand miners have completely obliterated at least two dozen Indonesian islands since 2005. Hauled off boatload by boatload, the sediment forming those islands ended up mostly in Singapore, which needs titanic amounts of sand to continue its program of artificially adding territory by reclaiming land from the sea. The city-state has created an extra fifty square miles in the past forty years and is still adding more, making it by far the world's largest sand importer. The demand has denuded beaches and riverbeds in neighboring countries to such an extent that Indonesia, Malaysia, Vietnam, and Cambodia have all restricted or completely banned exports of sand to Singapore.

The sand underneath the water isn't safe, either. Sand miners are increasingly turning to the seafloor,[25] vacuuming up millions of tons with dredges the size of aircraft carriers. One-third of all aggregate used in construction in London and southern England comes from beneath[26] the United Kingdom's offshore waters. Japan relies on sea sand even more heavily, pulling up around 40 million cubic meters from the ocean floor each year.[27] That's enough to fill up the Houston Astrodome thirty-three times.

Hauling all those grains from the seafloor tears up the habitat of bottom-dwelling creatures and organisms. The churned-up sediment clouds the water, suffocating fish and blocking the sunlight that sustains underwater vegetation.[28] The dredging ships dump grains too small to be useful, creating further waterborne dust plumes that can affect aquatic life far from the original site.[29]

Dredging of ocean sand has also damaged coral reefs in Florida and many other places, and threatens important mangrove forests, sea grass beds, and endangered species such as freshwater dolphins[30] and the Royal Turtle.[31] One round of dredging may not be significant, but the cumulative effect of several can be. Large-scale ocean sand mining is new enough that there hasn't been a lot of research on it, meaning that no one knows for sure what the long-term environmental impacts will be. We're sure to find out in the coming years, however, given how fast the practice is expanding.

Sand mining is also damaging lands and livelihoods far from any coast. The fracking boom in the United States has created a voracious hunger for what's known as "frac sand." Fracking is the deeply controversial method of extracting oil and gas from shale rock formations by breaking—that is, fracturing—the subterranean stone by blasting it with a high-pressure mix of water, chemicals, and a particular type of especially hard, rounded sand grains. It happens that there are huge deposits of just that kind of sand in Minnesota and Wisconsin. Result: the fracking rush in North Dakota has sparked a frac sand rush in the Upper Midwest. Thousands of acres of fields and forests have been stripped away so that miners can get their hands on those rare grains.

Colossal amounts of more ordinary construction sand is dredged up from riverbeds or dug from nearby floodplains. In central California, floodplain sand mining has diverted river waters into dead-end detours and deep pits that have proven fatal traps for salmon.[32] In northern Australia, floodplains that are home to the world's biggest collection of rare carnivorous plants are being wiped out by sand mining.[33]

Dredging sand from riverbeds, as from seabeds, can destroy habitat and muddy waters to a lethal degree for anything living in the water. Kenyan officials shut down all river sand mines in one western province in 2013 because of the environmental damage they were causing. In Sri Lanka,[34] sand extraction has left some riverbeds so deeply lowered that seawater intrudes into them, damaging drinking water supplies. India's Supreme Court warned in 2011 that "the alarming rate of unrestricted sand mining" was disrupting riparian ecosystems all over the country, with fatal consequences for fish and other aquatic organisms and "disaster" for many bird species.[35]

In Vietnam, researchers with the World Wildlife Federation believe sand mining on the Mekong River is a key reason the 15,000-square-mile Mekong Delta—home to 20 million people and source of half of all the country's food and much of the rice that feeds the rest of Southeast Asia—is gradually disappearing. The ocean is overtaking the equivalent of one and a half football fields of this crucial region's land every day. Already, thousands of acres of rice farms have been lost, and at least 1,200 families have had to be relocated from their coastal homes. All this is caused partly by climate-change-induced sea level rise, and partly by direct human intervention. For centuries, the delta has been replenished by sediment carried down from the mountains of Central Asia by the Mekong River. But in recent years, in each of the several countries along its course, miners have begun pulling huge quantities of sand from the riverbed to use for the construction of Southeast Asia's surging cities. Nearly 50 million tons of sand are being extracted annually—enough to cover the city of Denver two inches deep. "The sediment flow has been halved," says Marc Goichot, a researcher with the World Wildlife Fund's Greater Mekong Programme. That means that while natural erosion of the delta continues, its natural replenishment does not. At this rate, nearly half the delta will be wiped out by the end of this century.

Sand extraction from rivers has also caused untold millions of dollars worth of damage to infrastructure around the world. The stirred-up sediment clogs up water supply equipment, and all the earth removed from riverbanks leaves the foundations of bridges exposed and unsupported. A 1998 study found that each ton of aggregate mined from the San Benito River on California's central coast caused $11 million in infrastructure damage—costs that are borne by taxpayers.[36] In many countries, sand miners have dug up so much ground that they have dangerously exposed the foundations of bridges and hillside buildings, putting them at risk of collapse.

That risk isn't just theoretical. In Taiwan in 2000, a bridge undermined by sand extraction gave way. The following year, the same thing happened to a bridge in Portugal just as a bus was passing over it; seventy people were killed.[37] Another bridge collapse in India in 2016 that killed twenty-six may have been caused by sand mining.

Sand mining can also directly harm people and their communities. Unprotected miners have died when sandpit walls collapsed on them. Fisherfolk from Cambodia to Sierra Leone are losing their livelihoods as sand mining decimates the populations of fish and other aquatic creatures they rely on. In some places, mining has made riverbanks collapse, taking out agricultural land and causing floods that have displaced whole families. In Vietnam in 2017 alone, so much soil slid into heavily mined rivers, taking with it the crops and homes of hundreds of families, that the government shut down sand extraction completely in two provinces. And in Houston, Texas, government officials say that sand mining in the nearby San Jacinto River—much of it illegal—seriously exacerbated flooding damage during 2017's Hurricane Harvey. It seems that sand miners stripped away so much vegetation along the river banks that huge amounts of silt were left exposed, and were then washed into the river by Harvey's rains. That silt then piled up in riparian bottlenecks and at the bottom of Lake Houston, the city's principal source of drinking water, causing them to overflow into nearby neighborhoods.

River-bottom sand also plays an important role in local water supplies. It acts like a sponge, catching the water as it flows past and percolating it down into underground aquifers. But when that sand has been stripped away, instead of being drawn underground, the water just keeps on moving to the sea, leaving aquifers to shrink. As result, there are parts of Italy and southern India where river sand mining has drastically depleted local drinking water supplies.[38] Elsewhere, the lack of water is killing crops. Researchers fear that sand mining in the Chaobai River, which feeds one of the main reservoirs supplying Beijing, may not only disrupt the river's ecosystem but also compromise the quality of the capital's drinking[39] water.

Even after the sand miners are done, the battered landscape they leave behind can be startlingly dangerous. In the United States and elsewhere, mining companies are generally required to restore the land to a certain extent after they are finished. But in less well-organized countries, miners leave behind deep open pits that fill with rainwater and trash, degenerating into swampy breeding grounds for disease-carrying mosquitoes. A number of children have reportedly drowned in such pits in recent years. In Sri Lanka and India, sand mining has destroyed crocodile habitats, sending the beasts closer to river shores, where they have killed at least half a dozen[40] people in the last ten years.

In response to all this destruction, governments around the world have tried, with varying levels of commitment, to regulate sand mining and to restrict the places and manner in which it is done. That in turn has spawned a booming worldwide black market in sand.

Illegal sand mining runs a wide gamut. At one end, it includes legitimate businesses overstepping the boundaries of their permits. In 2003, for instance, California filed a lawsuit[41] against Hanson

Aggregates, a global mining outfit, for unauthorized dredging of sand from the San Francisco Bay. "These sand pirates have enriched themselves by stealing from the state and ripping off taxpayers," the state's attorney general declared at the time. Hanson eventually settled, paying the state $42 million.

At the other extreme are outright criminals, from petty thieves to well-organized gangs willing to kill to protect their sand business. In 2015, New York state authorities slapped a $700,000 fine on a Long Island contractor who had illegally gouged thousands of tons of sand from a 4.5-acre patch of land near the town of Holtsville and then refilled the pit with toxic waste. These "scoop and fill" operations have become common as the area's legitimate sources of sand have been increasingly depleted, according to the New York State Department of Environmental Conservation.[42]

In other countries, the black market takes more dramatic forms. One of Israel's most notorious gangsters, a man allegedly involved in a spate of recent car bombings, got his start stealing sand from public beaches. In Morocco, fully half the sand used for construction is estimated to be mined illegally; whole stretches of beach in that country are disappearing.[43] In Kenya illegal sand miners reportedly coax children into dropping out of school to come work for them. South Africa has set up a dedicated squad of police dubbed the Green Scorpions to combat illegal sand mining. Sometimes the criminal sand trade crosses borders. Dozens of Malaysian officials were charged in 2010 with accepting bribes and sexual favors in exchange for allowing illegally mined sand to be smuggled into Singapore.

Like any big-money black market, sand also generates violence. People have been shot, stabbed, beaten, tortured, and imprisoned over sand mining in countries around the world—some for trying to stop the environmental damage, some in battles over control of the land, and some caught in the cross fire. In Cambodia, police have jailed environmental activists who boarded river dredges to protest against illegal mining. In Ghana, security forces have opened fire on rowdy demonstrations against local sand miners. In China, a dozen members of a sand mining gang were sent to prison in 2015 after battling with knives in front of a police station. In Indonesia in 2016, an activist was beaten into a coma, and another tortured and stabbed to death, by the sand miners they were trying to stop. In Kenya, at least nine people have been killed—including a policeman hacked to death with machetes—in battles between farmers and sand miners in recent years.

To understand how the demand for sand can get so intense, and how it can spawn such destruction, in 2015 I started looking into the illegal sand trade in India. India is ground zero of the global sand crisis, the home of the blackest of the world's black markets in the stuff. The *Times of India*[44] estimates that the illicit sand trade is worth some $2.3 billion a year. Battles among and against "sand mafias" there have reportedly killed hundreds of people in recent years—including police officers, government officials, and ordinary people who get in their way. I had an unexpected and somewhat stressful encounter with some of these mafiosi not long ago, while I was investigating a murder so brazen it was hard to believe it had happened.

A little after eleven A.M. on July 31, 2013, the sun was beating down on the low, modest residential buildings lining a back street in the Indian farming village of Raipur Khadar, southeast of New Delhi. Faint smells of cooking spices, dust, and sewage seasoned the air.[45]

In the back room of a two-story brick-and-plaster house, Paleram Chauhan, a fifty-two-year-old vegetable farmer, was napping after an early lunch. In the next room, his wife and daughter-in-law were cleaning up while Paleram's son Ravindra played with his three-year-old nephew.

Suddenly gunshots thundered through the house. Preeti Chauhan, Paleram's daughter-in-law, rushed into Paleram's room, Ravindra right behind her. Through the open back door, they saw two men with white kerchiefs covering their lower faces. One was holding a pistol. The men piled onto a motorcycle driven by a third and roared away.

Paleram lay on his bed, blood bubbling out of his stomach, neck, and head. He stared at Preeti, trying to speak, but no sound came from his mouth. Ravindra borrowed a neighbor's car and rushed his father to a hospital, but it was too late. Paleram was dead on arrival.

Despite the masks, the family had no doubts about who was behind the killing. For ten years Paleram had been campaigning to get local authorities to shut down a powerful gang of criminals headquartered in Raipur Khadar. The "mafia," as people called them, had for years been robbing the village of one of its most precious resources: sand.

The area around Raipur Khadar used to be mostly agricultural—wheat and vegetables growing in the Yamuna River floodplain. But Delhi, India's capital and the world's second biggest city with a population topping 25 million, is less than an hour's drive north, and it is encroaching fast. Driving down a new six-lane expressway that cuts through Gautam Budh Nagar, the district in which Raipur Khadar sits, I passed construction site after construction site, new glass and cement towers sprouting skyward like the opening credits from *Game of Thrones* made real across miles of Indian countryside. Besides countless generic shopping malls, apartment blocks, and office towers, a 5,000-acre

"Sports City" was under construction, including several stadiums and a Formula 1 racetrack.

The building boom got in gear in the mid-2000s, and so did the sand mafias. "There was some illegal sand mining before," said Dushynt Nagar, the head of a local farmers' rights organization, "but not at a scale where land was getting stolen or people were getting killed."

The Chauhan family has lived in the area for centuries, Paleram's son Aakash told me. He's young and slim, with wide brown eyes and receding black hair, wearing jeans, a gray sweatshirt, and flip-flops. We were sitting on plastic chairs set on the bare concrete floor of the family's living room, just a few yards from where his father was killed.

The family owns about ten acres of land and shares some two hundred acres of communal land with the village—or used to. About ten years earlier, a group of local musclemen, as Aakash calls them, led by Rajpal Chauhan (no relation—it's a common surname) and his three sons, seized control of the communal land. They stripped away its topsoil and started digging up the sand built up by centuries of the Yamuna's floods. To make matters worse, the dust kicked up by the operation stunted the growth of surrounding crops.

As a member of the village panchayat, or governing council, Paleram took the lead in a campaign to get the sand mine shut down. It should have been pretty straightforward. Aside from stealing the village's land, sand mining is not permitted in the Raipur Khadar area at all, because it's close to a bird sanctuary. And the government knows it's happening: in 2013 a fact-finding team from the federal Ministry of Environment and Forests found[46] "rampant, unscientific, and illegal mining" all over Gautam Budh Nagar.

Nonetheless, Paleram and other villagers couldn't find anyone willing to help. They petitioned police, government officials, and courts for years—and nothing happened. Conventional wisdom is that many local authorities accept bribes from the sand miners to stay out of their business—and not infrequently are involved in the business themselves.

For those who don't take the carrot of a bribe, the mafias aren't shy about using a stick. "We do conduct raids on the illegal sand miners," said Navin Das, the official in charge of mining in Gautam Budh Nagar. "But it's very difficult because we get attacked and shot at."

Since 2014, Indian sand miners have killed at least seventy people, including seven police officers and more than half a dozen government officials and whistle-blowers. Many more have been injured, including journalists. Just a few months after my trip to India in 2015, an assault by illegal sand miners put a television reporter in the hospital. Shortly after that, another journalist investigating illegal sand mining was burned to death.

Rajpal and his sons warned Paleram and his family, as well as other villagers, to stop making trouble for them—or else. Aakash knows one of the sons, Sonu, from when they were kids in school together. "He used to be a decent guy," Aakash said. "But when he got into the sand business and started making fast money, he developed a criminal mentality and became very aggressive." Rather than backing down, the villagers filed reports of the threats with local courts. Finally, in the spring of 2013, police arrested Sonu and impounded some of his outfit's trucks. He quickly posted bail.

One morning soon afterward, Paleram rode his bicycle out to his fields, which are right next to the sand mine, and ran into Sonu.

"Sonu said, 'It's your fault I was in jail,'" according to Aakash. "He told my father to drop the issue." Instead Paleram complained to the police again.

Just a few days later, Paleram was shot dead.

Sonu, his brother Kuldeep, and his father, Rajpal, were arrested for the killing. All of them were soon out on bail. Aakash sees them around sometimes. "It's a small village," he said.

A akash agreed to show me and my interpreter, Kumar Sambhav, the village lands where the mafia had taken over. We'd rented a car in Delhi that morning, and Aakash directed our driver to the site. It was hard to miss: right across the road from the village center is an expanse of torn-up land pocked with craters ten and twenty feet deep, stippled with house-sized piles of sand and rock. We drove in, picking our way carefully along the rutted dirt track running through the mine. Here and there trucks and earth-moving machines rumbled around, and clusters of men, at least fifty in all, were smashing up rocks with hammers and loading trucks with shovelfuls of sand. They stopped to stare at our car as we trundled past. Aakash cautiously pointed out a tall, heavyset guy in jeans and a collared shirt: Sonu.

A short while later, deep inside the site, we got out of the car so I could snap pictures of a particularly huge crater. After a few minutes Aakash spotted four men, three of them carrying shovels, striding purposefully toward us. "Sonu is coming," he muttered.

We started making our way back to the car, trying to look unhurried. We were too slow. "Motherfucker!" Sonu, now just a few yards away, barked at Aakash. "What are you doing here?"

Aakash kept silent. Sambhav mumbled something to the effect

that we were just tourists as we all climbed into the car. "I'll give you sisterfuckers a tour," Sonu said. He yanked open our driver's door and ordered him out. The driver obeyed, obliging the rest of us to follow. Aakash, wisely, stayed put.

"We're journalists," Sambhav said. "We're here to see how the sand mining is going." (This conversation was all in Hindi; Sambhav translated for me afterward.)

"Mining?" Sonu said. "We are not doing any mining. What did you see?"

"We saw whatever we saw. And now we're leaving."

"No, you're not," Sonu said.

The exchange continued along those lines for a couple of increasingly tense minutes, until one of Sonu's goons pointed out the presence of a foreigner—me. This gave Sonu and his crew pause. It's bitterly unfair, but harming a Westerner like me could bring them a lot more trouble than going after a local like Aakash. There was a confused momentary stalemate. We grabbed the opportunity to jump back in the car and take off. Sonu, glaring, watched us go.

At the time of this writing, the case against Sonu and his relatives was still grinding its way through India's sluggish courts. The outlook wasn't great. "In our system you can easily buy anything with money—witnesses, police, administrative officials," a legal professional close to the case told me, on condition of anonymity. "And those guys have a lot of money from the mining business."

Aakash keeps in touch with police investigators and has tried to get India's National Human Rights Commission to take an interest in his father's murder. His mother pleads with him to drop the whole thing, especially since her other son, Aakash's brother, Ravindra—who was to have been the main witness in the case—was found dead by some railroad tracks last year. He was apparently run over by a train. No one is quite sure how that happened.

Elsewhere around India, many others are trying in many ways to get sand mining under control. The National Green Tribunal, a sort of federal court for environmental matters, has opened its doors to any citizen to file a complaint about illegal sand mining. Villagers have organized demonstrations and blocked roads to stop sand truck traffic. Nearly every day some local or state official declares their determination to combat sand mining. They have impounded trucks, levied fines, and arrested people. Police have even started using drones to spot unauthorized mining sites.

But India is a vast country of more than 1 billion people. It hides hundreds, most likely thousands, of illegal sand mining operations. Corruption and violence will stymie many of even the best-intentioned attempts to crack down on them.

And it's not just India. There is large-scale illegal sand extraction going on in dozens of countries. One way or another, sand is mined in almost every country on Earth. India is only the most extreme manifestation of a slow-building crisis that affects the whole world.

At root, it's an issue of supply and demand. The supply of sand that can be mined sustainably is finite. But the demand for it is not.

Every day the world's population is growing. More and more people in India—and everywhere else—want decent housing to live in, offices and factories to work in, malls to shop in, and roads to connect them. Economic development as it has historically been understood requires concrete and glass. It requires sand.

People have used sand for millennia. But only in the twentieth century, with the advent of modernity, did it become indispensable to the Western world. In the twenty-first century, in our digital, globalized era, sand has become indispensable to almost everyone. A century ago, a few hundred million people lived in a way that required a great deal of sand—residing and working in concrete structures, traveling on asphalt roads, with glass windows everywhere. Today, billions live that way, and the number is growing by the day. Sand has become one of the twenty-first century's most sought-after commodities, sparking violence and destruction around the world.

How did we get here? How did we become so dependent on such a simple material? How can we possibly be using so much of it? And what does our dependence mean for the planet, and for our future?

PART I

How Sand Built the Twentieth Century's Industrialized World

Nothing is built on stone;
all is built on sand, but we must
build as if the sand were stone.

—JORGE LUIS BORGES,
In Praise of Darkness

CHAPTER 2

The Skeleton of Cities

At 5:12 in the morning of April 18, 1906, a titanic earthquake sledgehammered the city of San Francisco. For almost a full minute, streets convulsed and buildings shivered and collapsed. People were killed by the dozens. "Of a sudden we had found ourselves staggering and reeling. . . . Then came the sickening swaying of the earth that threw us flat upon our faces," recalled one eyewitness. "Then it seemed as though my head were split with the roar that crashed into my ears. Big buildings were crumbling as one might crush a biscuit in one's hand. Ahead of me a great cornice crushed a man as if he were a maggot—a laborer in overalls on his way to the Union Iron Works with a dinner pail on his arm."[1]

Horrific as the quake was, even worse was yet to come. The temblor ruptured gas mains, sparking massive fires that raged for three solid days, ravaging thousands of buildings and incinerating hundreds of people.

After the flames finally burned themselves out, a curious sight emerged at the intersection of Mission and Thirteenth streets. Amid the heaps of charred beams and broken brick, a single building still stood. It was an unassuming half-finished warehouse owned by Bekins Van and Storage. It survived because it had been built with a controversial new material called reinforced concrete. The countless tiny sand soldiers embedded in its concrete walls and floors had given it the strength to resist the flames. Hardly anyone realized it at the time, but this otherwise unremarkable warehouse signaled a watershed moment in the history of architecture, construction, and humankind itself.

Concrete is an invention as transformative as fire or electricity. It has changed where and how billions of people live, work, and move around. Concrete is the skeleton of the modern world, the scaffold on which so much else is built. It gives us the power to dam enormous rivers, erect buildings of Olympian height, and travel to all but the remotest corners of the world with an ease that would astonish our ancestors. Measured by the number of lives it touches, concrete is easily the most important man-made material ever invented.

This world-transforming substance is composed mainly of the simplest, most commonplace ingredients: gravel and sand. Concrete, in fact, is the primary driver of the global sand crisis; we use far more sand to make concrete than for any other purpose. Billions of tons of sand and gravel are unearthed every year and pressed into service to form shopping malls, freeways, dams, and airports. The whole substrate of the world we live in rests on the shoulders of that vast infantry of miniature stones.

All of which is even more amazing when you consider that only a little over a century ago, we barely used concrete at all.

Let's clear up one thing right away: Cement is not the same thing as concrete. Cement is an *ingredient* of concrete. It's the glue that binds the gravel and sand together. Cements (there are many forms) are typically made by crushing up clay, lime, and other minerals, firing them in a kiln at temperatures up to 2,700 degrees, then milling the result into a silky-fine gray powder. Mix that powder with water and you get a paste. The paste doesn't simply dry, like mud; it "cures," meaning the powder's molecules bond together via a process called hydration, its chemical components gripping each other ever tighter, making the resulting substance extremely strong. Reinforced with a platoon of sand, that paste thickens into mortar, the stuff used to hold bricks together.

Concrete is made by adding "aggregate"—sand and gravel—to the mix of cement and water. Typical concrete is about 75 percent aggregate, 15 percent water, and 10 percent cement. Combine those materials, and the result is a gloopy gray liquid that can be poured into virtually any shape. As the cement cures, it binds to the aggregate, locking the grains together like a zillion tiny bricks and hardening the whole mess into solid artificial stone.

Though concrete is the quintessential modern building material, people in several places over the centuries have stumbled on the trick of making it. The Mayans, who flourished 2,000 years ago in what is now southern Mexico, Guatemala, and Belize, made crude concrete beams to support some of their buildings.[2] The Greeks used cement mortars. (Some scholars believe the ancient Egyptians used a form of concrete in the building of pyramids, though most disagree. The Egyptians almost certainly did use sand, though, to help their bronze saws cut through stone for their monuments,[3] likely including the pyramids. Sand, in fact, has been used for construction since at least 7000 BCE, by ancient peoples who mixed it with mud to make crude bricks.) But by far the most enthusiastic and technically sophisticated users of concrete in the ancient world were the Romans.

It's not clear exactly when or how the Romans figured out the secret of concrete making. The task was certainly made easier by their lucky discovery of a type of naturally occurring cement in Pozzuoli, near Naples.[4] The earliest known Roman concrete dates back to the third century BCE.[5] "The Romans recognized the potential of this material and would use it with gusto throughout their empire until its fall in the fifth century," writes author Robert Courland in *Concrete Planet*.[6] "They systematized its production and application and were the first people to utilize concrete as we do today: putting it into large molds to create a strong monolithic architectural unit." (The Romans didn't use the term *concrete*, though it is derived from the Latin *concretus*, meaning brought together or congealed.)

Roman engineers developed sophisticated techniques to improve on basic concrete. Concrete shrinks as it hardens, which can cause it to crack. Water seeping into the cracks expands when it freezes, widening those cracks and further weakening the concrete. Adding horsehair helped with shrinkage, the Romans found, and putting a bit of blood or animal fat in the mix helped the concrete withstand the effects of freezing water.[7]

The Romans built houses, shops, public buildings, and baths from concrete. The breakwaters, towers, and other structures that made up the colossal man-made harbor of Caesarea,[8] in what is now Israel, were built with concrete, as was the foundation of the Colosseum, along with countless bridges and aqueducts[9] across the empire. Most famously, Rome's Pantheon, built nearly 2,000 years ago, is roofed with a spectacular concrete dome—still the biggest concrete structure without reinforcing steel in the world.

Like so much other knowledge the Romans had accumulated, though, the science and technology of concrete faded from memory as the empire slowly crumbled over the centuries that followed. "Perhaps the material was lost because it was industrial in nature and needed an industrial empire to support it," writes scientist and engineer Mark Miodownik in *Stuff Matters*. "Perhaps it was lost because it was not associated with a particular skill or craft, such as ironmongery, stonemasonry, or carpentry, and so was not handed down as a family trade."[10] Whatever the reasons, the result was striking: "There were no concrete structures built for more than a thousand years after the Romans stopped making it," notes Miodownik.

It was the British, those indefatigable experimenters (and long-ago Roman subjects), who started bringing concrete back. In the 1750s, an English engineer named John Smeaton, while tinkering with various binding agents to hold together the granite blocks he was using to build a lighthouse off the coast of Plymouth, came up with an excellent formula for hydraulic (water-using) cement. (You can add other materials like gypsum, and play around with burning temperature and grain size, to change the properties of cement.[11] Today, there are hundreds of formulas for making cement tailored to specific weather conditions, project types, and other variables.)

Others continued tweaking the mixture, which came to be called Roman cement. By the early 1800s, hydraulic cement was sufficiently trusted to be used in the construction of a tunnel for horse-drawn carriages under the Thames River.[12] The tunnel was later adapted for trains, and reconstituted in 2010 as a museum.[13]

In 1824, a forty-four-year-old English bricklayer named Joseph Aspdin was granted a patent for his own cement formula. It was a mixture of powdered limestone and clay, fired at high temperatures, which he dubbed Portland cement, since its color was similar to the famous limestone from the Isle of Portland[14] in southern England. Aspdin had been trying for some time; hard up for expensive materials, he was twice charged with stealing limestone from paved roads. His was just one of many patents given in those years to inventors of various cement formulas, but his took off. That was partly because it was stronger and more durable than the competition, and partly because Aspdin's son William seriously exaggerated its quality in his marketing pitch.[15] Nonetheless, it has become the industry standard; today, 95 percent of the roughly 83 million tons of cement manufactured in America is Portland cement.[16]

Various tinkerers were intrigued by the crude concrete you could make by mixing sand and gravel with Aspdin's cement. In the early 1800s, an artist named James Pulham started making vases, sculptures, and architectural adornments out of concrete. Others tried it for architectural purposes. "Although largely ignored by most people during much of the nineteenth century, the idea of using concrete to cast walls and floors to make houses was an appealing challenge for a few brave souls active in the cement industry," writes Courland. "Perhaps a dozen concrete houses were built in England in the 1850s, and a few still remain."[17]

The problem with concrete is that it has tremendous compressive strength, which means it can stand up to great pressure without breaking, but it has little tensile strength, which means it can't bend much without shattering. That limited its usefulness. By the mid-1800s, inventors and entrepreneurs were looking for ways to increase concrete's tensile strength. The most promising approach was to embed it with iron, essentially giving it an inner skeleton that absorbs the stress from bending pressure, keeping the concrete from fatally cracking.[18]

A French farmer came up with the unlikely-seeming idea of using concrete reinforced with iron bars to build a boat. The thing did actually float—for a little while, anyway. Then it sprang a leak and promptly sank to the bottom of the farmer's pond. In 1867, a gardener named Joseph or Jacques (accounts differ) Monier, another Frenchman, wanted stronger tubs for large plants than the standard fired-clay ones. He came up with a system of reinforcing concrete with loops of metal wire.[19]

This was a crucial breakthrough. On its own, concrete is basically artificial stone. Reinforced with iron or steel, though, it becomes a building material unlike anything found in nature, one that combines the strengths of both metal and stone. That's what makes it so useful for so many purposes.[20]

Builders in Europe and the Americas dabbled with the new material.[21] The first home built with reinforced concrete went up in Rye Brook, New York, in the early 1870s, the project of an engineer named William Ward. It's still there. At the time, it was the world's largest reinforced concrete structure.

It was right around this time that a young man named Ernest L. Ransome set out from his home in Ipswich, England, to seek his fortune in booming, bawdy San Francisco. Ransome was a scion of a family of ironworkers and engineers that had helped develop products from lawn mowers to ball bearings. Ransome's father, Frederick, branched out into making and selling artificial stone, and developed his own cement mixture. Ernest started apprenticing in his father's factory in 1859 at age seven. At that time, as he later wrote, "the concrete industry was in its infancy, and was confined largely to the manufacture of artificial stone for ornamental purposes."[22]

Ransome, a trim and stern-faced fellow, arrived in San Francisco in the early 1870s. It was an excellent place and time for an ambitious, inventive type. Grown rich from the Gold Rush, the city was by then a hub for the new Silver Rush in nearby Nevada, and a base for moguls of the mining, manufacturing, and railroad industries. It was growing fast; the population quadrupled between 1860 and 1880 to nearly a quarter of a million.[23] Ransome found a job at a company that produced concrete blocks for paving stones and architectural decorations,[24] and talked his colleagues into switching over to his father's brand of cement. Within a few years, he left to start up his own outfit. He sold concrete vases and cement components (he eventually abandoned his father's brand for the standard Portland cement), and in his spare time noodled around trying to develop new reinforcing techniques that would make stronger, more durable, more versatile concrete.

In the early 1880s, San Francisco city authorities decided that the standard wooden sidewalks weren't strong enough to cope with the growing numbers of pedestrians that were pounding up and down them every day. They began replacing the old walkways with sturdier ones made of concrete. This of course was great for the concrete makers' business. The *San Francisco Chronicle* reported in 1885 that sales were surging as "artificial stone for sidewalks and basements is coming into general use in nearly all the larger towns on the coast."[25]

One forward-thinking local contractor built some of these sidewalks using a technique, patented by an American inventor named Thaddeus Hyatt,[26] of reinforcing the concrete with embedded iron bars. Impressed with the results, Ransome set about experimenting with variations on Hyatt's method and soon came up with a historic innovation. He took two-inch-thick square iron bars, attached their ends to an adapted cement mixer he set up in his backyard, and twisted the bars, like a towel being wrung. The twisted bars gripped the concrete more firmly all along their length, and the process of twisting them also increased their tensile strength. It was the first version of the now-standard steel rebar used in reinforced concrete structures around the world.

Nonetheless, as Ransome recalled a few years later, convincing his peers wasn't easy. "When I presented my new invention to the technical society of California, I was simply laughed down, the consensus of opinion being that I injured the iron," he writes in his prosaically titled book, *Reinforced Concrete Buildings*. It took many tests before he began to win converts.[27] Ransome patented the system in 1884, the same year he built the first large commercial structure made with reinforced concrete, a warehouse in San Francisco for the Arctic Oil Company. He followed that with the Alvord Lake Bridge, an arched pedestrian tunnel under the main thoroughfare running through Golden Gate Park, and two important buildings for the campus of the new Stanford University, south of the city in Palo Alto.

Reinforced concrete kept proving itself, and Ransome's business grew rapidly. He became the nation's foremost evangelist of concrete. He was awarded patents[28] for a range of additional processes and machines, and began leasing out his system for use far and wide. One of the reasons for his success was that Ransome was a stickler when it came to sand. There are gradations of quality even among common construction sand, and Ransome would accept only the finest into his service. "Next to the cement, the sand is the most important factor in determining the strength of the concrete," he told would-be builders in his book. "It is well understood by skilled concrete men that the best grade of sand is clean, sharp, and well graded from fine to coarse."[29]

Meanwhile, the price of steel was plummeting, thanks to rapidly advancing production methods and the discovery of titanic deposits of iron, steel's basic raw material, in Minnesota. Those lower prices made it feasible to replace iron rebar in concrete with steel, making the concrete even stronger. Cement was getting cheaper, too, giving concrete an economic edge over steel and masonry buildings. The upstart material made headlines around the world in 1901, when contractors using Ransome's system put up

Cincinnati's sixteen-story Ingalls Building, by far the tallest concrete structure on the planet and one that nearly matched the height of the biggest skyscrapers then in existence.

Still, by 1906 there were very few reinforced concrete buildings in California. That was largely thanks to bitter opposition from powerful building trade unions, especially on Ransome's home turf of San Francisco.[30] Bricklayers, stonemasons, and others, correctly seeing in concrete a mortal threat to their professions, denounced it as unproven and unsafe. Just a few months before the quake, a group of bricklayers and steelworkers in Los Angeles tried to convince the city council to forbid the construction of any more concrete buildings[31] within municipal limits.

The tradesmen also made a case against concrete on the grounds that it was plain ugly. An article in *The Brickbuilder*, a monthly trade publication, complained in May of 1906 that "a city of the dull grayness of concrete would defy all laws of beauty. . . . Concrete does not lend itself architecturally to anything that appeals to the eye. Let us pause a moment before we transform our city into such hideousness as has been suggested by concrete engineers and others interested in its introduction."[32]

Concrete, however, kept gaining ground. In retrospect, the process to a certain extent resembled the rise of computers many decades later. At first, people could see that this new technology was promising, but who knew if it would actually work better than the tried and trusted old ways? Why risk your business on some new-fangled invention when your trusty paper ledgers, or your dependable bricks, did the job just fine? For quite a few years it was only the early adopters—the inventors, the hackers, the hobbyists—who played around with the new thing in its early, crude forms, figuring out how it could be used. But gradually concrete, like the computer, became more refined, dependable, and easier to use, until it reached a point where practically anyone could work with it.

There was no single point at which concrete definitively eclipsed other building methods. But the fact of the survival of the Bekins warehouse, along with the many other concrete foundations, floors, and full-scale buildings that stood up well to the 1906 earthquake and subsequent fire, was a watershed. (The warehouse was in such good shape that the company turned it into a shelter for newly homeless locals.)[33] The concrete industry certainly thought so, and wasn't shy about using photos of the rubble to promote their cause. "The American cement industry has grown up through a mass of prejudice, the last vestige of which was overthrown and buried by the splendid showing made by concrete in the San Francisco earthquake and fire," declared the June 1906 issue of *Cement and Engineering News*.[34]

Trade press editors weren't the only ones convinced. Captain John Sewell of the Army Corps of Engineers, one of three authors of a 1907 report commissioned by the US Geological Survey on the San Francisco earthquake's damage, declared that the "great utility of reinforced concrete in earthquake shocks can not be denied" and that a "solid monolithic concrete structure of any sort is secure against serious damage in any earthquake country," unless "it should happen to lie across the line of the slip [seismic fault]." He also decried the "opposition of the bricklayers' union and similar organizations" that had "prevented the use of reinforced concrete in San Francisco for all parts of buildings. This action of the labor unions will cost the city a good deal, and, should it be continued, will cost a great deal more in the future."[35]

In *Concrete Planet,* Courland contends that Sewell and the other authors of the USGS report were "biased in favor of reinforced concrete construction and against masonry building," noting that one of them later became president of the National Association of Cement Users. Indeed, several reinforced concrete buildings in San Francisco were seriously thrashed by the quake, while some brick buildings came through just fine—facts that were ignored or down-played by the USGS investigators.[36]

It didn't matter. Concrete won the public relations battle. An article in the *San Francisco Chronicle* a few weeks after the fire gushed that "these buildings and parts of buildings passed the or-deal of the earthquake practically uninjured . . . re-enforced concrete roofs and floors passed triumphantly through the earth-quake." The newspaper concluded: "We now have re-enforced concrete, in great measure perfected and proved for our use. With it we can . . . build comparatively light and even graceful and hand-some structures that will have the bearing strength of natural stone, the tensile strength of steel to resist the disrupting influence of shocks, much of the artistic effect of carved stone, and a lasting and fire-resisting quality which will surpass them all."[37]

San Francisco building codes, however, still forbade the use of concrete in high, load-bearing walls. Ransome and his fans wanted that provision changed, but traditional tradesmen saw it as their last line of defense. The urgent need to start rebuilding the city gave impetus to concrete's case. Some 225,000 people were left homeless by the quake, more than half the city's population. (A *Los Angeles Times* article on the dispute added that labor shortages were an-other issue slowing down construction. Things were so dire that one "William Maxwell of the Pacific Wrecking Company has been forced to employ Japanese, paying them white man's wages.")[38]

Two months after the quake, the San Francisco board of super-visors held a meeting to discuss whether to change the code. So many would-be speakers on both sides of the argument showed up that one of the supervisors complained that hearing them all "would take a year." In the end, the anti-concrete faction lost. The board allowed concrete construction to go ahead.

The bricklayers didn't give up, though. The following year the union banned its members from working on buildings using concrete, and threatened to boycott "every other branch of the building industry connected with them," reported the *San Francisco Chronicle*.[39] But by then the war was already lost. "There is scarcely a block in the down-town burned district but will not soon boast of at least one reinforced concrete building, for they are to be on every hand seen in various stages of construction," reported a local newspaper in 1907.[40] By 1910, the city had issued permits for 132 new reinforced concrete buildings. Moreover, nearly all new steel-frame buildings built after the fire included concrete floors. "There were still obstacles to building with reinforced concrete as late as 1911, but these only slowed down the use of concrete," writes architectural historian Sara Wermiel. "The floodgates were open."[41]

A few months after the earthquake, Thomas Edison—the Steve Jobs of his day, inventor of the lightbulb, the phonograph, and much else—gave an after-dinner speech to a crowd of New York dignitaries assembled in his honor. Someone asked him what his next miraculous invention would be. "Concrete houses," replied Edison. Imagine, he told his audience: a home immune to fire, termites, mildew, and natural disasters.

Edison had been a believer in concrete for years. He had built a huge cement plant in New Jersey in 1899, and racked up a number of patents related to concrete and cement. In the wake of the earthquake, he became a full-fledged evangelist.

"It requires only one part of hydraulic Portland cement, mixed with three parts of sand and five parts of gravel . . . to make concrete as hard as adamant. I can put up a concrete building for about half the cost of a brick one," Edison told a reporter for the *San Francisco Call* soon after the New York dinner. "I not only propose to construct the outside walls of my house with cement, but [also] the walls forming the interior divisions, the stairs, the mantels and fireplaces." To top it off, he aimed to decorate the house with concrete "scrolls and flowered designs."[42] Later, he promised to bring to market concrete furniture "that will make it possible for the laboring man to put furniture in his home more artistic and more durable than is now to be found in the palatial residences in Paris or along the Rhine."[43] He could and would make practically anything out of concrete, Edison insisted—even pianos.

Such was the prestige, even glamour, that concrete enjoyed after its literal trial by fire in San Francisco. Nowadays, when we think of concrete (if we think of it at all), we tend to associate it with ugliness and oppression—the featureless walls of prisons, the dreary, dehumanizing concrete jungle. But once upon a time it seemed almost miraculous, a manifestation of progress, the harnessing of the earth's most basic materials to fulfill mankind's most exalted ambitions. Edison's home-building project fizzled out, and his concrete pianos never played a concert, but that didn't slow concrete's march to world domination.

"The rapid growth of reinforced concrete in public favor has been little short of marvelous. It is now used for nearly every form of structure for which timber, steel, or masonry is suitable," declared *Scientific American*[44] in 1906. Around the world, concrete office buildings, apartment blocks, hotels, dams, roads, statues, even ships were being built by the hundreds.[45] "Are there no limits to the conquests of concrete?" marveled the *Los Angeles Herald* in 1908. "Every day this new-old building material, as hard as stone, as strong as steel, almost as cheap as lumber, and as plastic as clay is put to some new use . . . Steel has been king for a long time. Concrete seems in a fair way to usurp the throne."[46]

Much like China and India today, the United States in those years was in the midst of twin population and urbanization booms. The country was adding an average of 1.5 million new citizens every year, and more and more Americans were moving to cities. The urban population nearly doubled between 1890 and 1910. By 1920, for the first time, more Americans lived in urban areas than on farms.[47] Increasingly, their homes, their workplaces, and the roads they traveled between them were made of concrete.

And the more concrete America used, the more sand it needed. Grains were hauled up in quantities never remotely seen before. In 1902, according to the US Geological Survey, the United States produced 452,000 metric tons of construction sand and gravel. Just seven years later, that amount had grown more than a hundredfold, to nearly 50 million tons.[48]

That sounds like a lot, until you learn that New York City's highways and skyscrapers, including the Empire State and Chrysler buildings, ate up more than 200 million tons of sand. Most of it was hauled in from Long Island, which still supplies a great deal of the city's needs. The abundant high-quality construction sands of the island is one reason Nassau County, just east of the borough of Queens, became such a popular suburb and vacation home site for New Yorkers. "The great hills of the north side of the county abound in sand that is excellent for building proposes," declared a 1912 *New York Times* article explaining the reasons behind the area's burgeoning growth. In addition, on the county's southern side, an "inexhaustible supply of beach sand" was being put to use to make concrete blocks "in nearly every community in the county."[49]

Sand has always been cheap, but when you're talking about quantities that large, there's a lot of money to be made. In 1919, a twenty-three-year-old eighth-grade dropout named Henry Crown and his brother Sol started a company with a borrowed $10,000 to supply sand and gravel to the contractors building Chicago. Sons of a Lithuanian immigrant sweatshop worker (née Krinsky), the Crown brothers would buy railcar loads of sand and deliver it by horse and wagon. Sol soon died of tuberculosis, leaving Henry in charge.

At the time, Chicago's population was exploding. It added half a million inhabitants between 1910 and 1920.[50] Supplying the building boom was a great business to be in. Crown's company, the Material Service Corporation, grew fast, buying its own sand and gravel pits, quarries, and processing plants. Within five years of founding the company, Crown was a millionaire. Later, Crown built custom-made barges equipped with pumps to suck sand from the bottom of Lake Michigan. His company's aggregate helped build Chicago's Loop railway and the Civic Opera House.

Crown expanded into real estate in a similarly big way: for several years he owned the Empire State Building. Material Service Corporation later became part of General Dynamics, America's biggest defense contractor. Still, Crown maintained a low-key attitude. "He would portray himself as a 'sand and gravel man' of limited education, veiling his moves and quietly consolidating his power," wrote *The New York Times* in his obituary. Crown died in 1990, the billionaire patriarch of one of America's wealthiest families.[51] The company that got him started is still a major aggregate outfit.

Concrete was well suited to the grandiose ambitions of the earliest twentieth century, when the Western world was at the peak of its power and hubris. Concrete made possible the Panama Canal, begun in 1903, which reshaped an entire nation's landscape and the world's shipping routes. It was used to make bunkers for millions of troops in World War I—a matter of such importance that the German military brought high-quality sand and gravel by barge from the Rhineland to the front lines, rather than relying on local supplies.[52] Concrete was used to build the titanic new factories cranking out automobiles and other industrial products all over the world. One million tons of it were deployed to anchor San Francisco's Golden Gate Bridge. The then-British colony of Hong Kong[53] produced so much concrete in the 1920s that sand supplies ran drastically short; thieves began stripping beaches and even digging up riverside graveyards, sparking violent clashes with villagers.

The capstone project of the era was the construction of the mighty Hoover Dam, at the time the biggest ever built. Enough sand and gravel to fill a train stretching 1,300 miles was mobilized to build this concrete monolith across the Colorado River. The process of harvesting, sorting, and hauling all that aggregate was a major engineering challenge all its own.

The job went to a California-based road building company owned by Henry J. Kaiser. Kaiser was at this point on his way to becoming one of America's wealthiest and most important industrialists; his deft handling of the sand and gravel supply for the dam was a key reputation builder. Kaiser and his aggregate expert Tom Price found a treasure trove of gravel and sand about six miles from the dam site, and there built one of the biggest aggregate plants the world had ever seen. In the facility's labyrinth of silos, conveyor belts, and storage containers, millions of tons of sand and gravel gouged from the earth with heavy equipment were sorted and sifted around the clock.

Sand was given special attention. When it came to making concrete, Price told an interviewer, "the secret of the important qualities of workability and uniformity are found to lie largely in the sand."[54] Once separated from the gravel, sand grains were further sorted by size in flotation tanks, in which, as a National Parks Service report later put it,[55] mechanical rakes would pull "a ribbon of wet sand out of the frothy water like some sort of prehistoric slime monster crawling out of the primordial ooze." The plant produced some 700 tons of aggregate per hour, its output loaded onto a specially built train to be hauled down to the dam.

Concrete has a way of leading to more concrete. The Hoover Dam created an enormous water supply called Lake Mead and also generated hydroelectricity. Together, those resources made it possible to build cities like Las Vegas and Phoenix in the middle of the desert—cities of concrete and glass and asphalt.

The spread of concrete also spawned whole new types of architecture. One of its earliest apostles was the American architect Frank Lloyd Wright,[56] who understood that concrete made possible entirely new forms. Take the inverted ziggurat of the Solomon R. Guggenheim Museum that Wright designed in New York. Wright created its fanciful geometry with "gun-placed concrete," aka gunite, a form of the compound made with more sand and less gravel than ordinary concrete, which allows it to be sprayed from a nozzle[57] directly onto a vertical surface. Try doing that with brick.

Wright's work paved, so to speak, the way for Walter Gropius's Bauhaus School, Le Corbusier's International school, and Richard Neutra's modernist creations. From Modernism grew Brutalism, the stark, angular, proudly concrete-heavy style that became popular after World War II. Today that term is often applied more broadly to the generic mode that has come to define so much of the visual landscape of our cities—the bluntly utilitarian look of near-identical factories and warehouses, the quadrangular shapes of institutional buildings and cheap apartment blocks, the coldly functional sweep of highway overpasses.

By the first decades of the twentieth century, sand and gravel in the form of concrete had become the ubiquitous building blocks of cities. Meanwhile, additional battalions of those little rock particles were being mobilized to create the roads that would knit cities together.

CHAPTER 3

Paved with Good Intentions

In the summer of 1919, a young US Army lieutenant colonel found himself stuck behind a desk at Maryland's Camp Meade, frustrated, depressed, and resentful. He had missed out on all the action in World War I, assigned instead of combat to overseeing a stateside training camp. He was bored with shuffling papers,[1] and he missed his wife and infant son, who were halfway across the country in Colorado. He was itching for something more exciting to do, especially if it might further his stalled career. So when word came around that the military was looking for volunteers to join a truck convoy that would cross the country from coast to coast, the twenty-eight-year-old officer—an ambitious West Point graduate by the name of Dwight Eisenhower—signed right up.[2]

"To those who have known only concrete and macadam highways of gentle grades and engineered curves, such a trip might seem humdrum," wrote the future president in his memoir *At Ease: Stories I Tell to Friends*. "In those days, we were not sure it could be accomplished at all. Nothing of the sort had ever been attempted."[3]

In today's America, so thoroughly defined, shaped, and organized around paved highways, it's hard to imagine just how few intercity roads there were, and how primitive they were, only a century ago. In 1904, the United States had a grand total of 141 miles of paved roads,[4] not counting city streets. Most of the rest were dirt tracks that devolved into mud in the winter and potholed, rutted obstacle courses in summer. Enormous stretches of land, especially in the West, didn't have *any* roads leading from one city to the next.

Crossing the continent in a motor vehicle was an exploit only a handful of hardy pioneers had attempted. A doctor from Vermont with the appropriately stirring name of Horatio Nelson Jackson was the first to succeed, slogging from San Francisco to New York in a two-cylinder, twenty-horsepower automobile. The trip took sixty-three days. A quartet of women led by New Jersey housewife Alice Huyler Ramsey made the same trek in the opposite direction a few years later, shaving four days off Jackson's time.[5]

By the time Eisenhower started packing for his cross-country road trip, the nation's highways were starting slowly to improve, thanks largely to the surging popularity of the automobile. Americans had bought over one million of these exhaust-spewing mechanical wonders by then, and were clamoring for better roads to drive their machines on. What was then called the War Department was also increasingly excited about the automobile's possibilities as a tool for combat. "The new vehicle, whose capacities had been well tested in training and combat support, offered a speed of movement and a mobility not restricted by rail schedules or routes," wrote Eisenhower. For the government, a cross-country convoy offered a chance to explore the military capabilities of cars and trucks, a solid publicity stunt, and a favor to the burgeoning auto industry.

The eighty-one-vehicle "motor truck train"—including trucks, motorcycles, ambulances, and field kitchens, accompanied by carloads of reporters and auto company officials—set off from Washington, DC, at 11:15 A.M. on July 7, 1919. Less than four hours later, a coupling broke on a kitchen trailer. That was only the first of many mechanical troubles that bedeviled the convoy. It advanced a grand total of forty-six miles that first day.

The worst problems, however, weren't with the vehicles, but what they had to travel on. Even the concrete roads that had been installed in parts of the more easterly states were often too narrow for the trucks, sending their tires off the pavement. Many had not been maintained since they were installed, leaving them in such ragged shape they could barely be driven on. The heavy trucks sometimes broke through the pavement and destroyed scores of too-flimsy bridges, forcing the vehicles to ford the occasional stream.[6]

That was the good news. In Illinois, the roads turned to dirt. "Practically no more pavement was encountered until reaching California," Eisenhower reported in his official notes. Motorcycle-riding scouts sped ahead of the convoy to find routes forward. For a long stretch between Utah and Nevada, Eisenhower noted with dismay, "the road is one succession of dust, ruts, pits, and holes."[7] Trucks got stuck in salt flats and stalled by sand drifts. At one point dozens of soldiers had to be harnessed to tow stranded trucks by hand.[8] Some days the convoy progressed only three miles. "There were moments when I thought neither the automobile, the bus, nor the truck had any future whatsoever," Eisenhower recalled.[9] When they finally reached San Francisco, they were greeted with speeches, a parade, and medals.

Along with pretty much every other officer on the journey, Eisenhower recommended to his superiors that somebody do something to improve America's roads. Many years later, he himself got to be that someone. In fact, he would launch the construction of what was for decades the most advanced and encompassing network of paved roads ever built: the US interstate highway system.

To build that continent-spanning network, the old general would call into service stratospheric quantities of construction sand. Every mile of the US interstate highway is made with some 15,000 tons of concrete.[10] Throw in the medians, overpasses, ramps, and road base, and all told, an estimated 1.5 billion tons[11] of gravel and sand went into making the national highway system. That's more than enough concrete to build a sidewalk reaching to the moon and back—twice.[12]

Laying down all that sand and gravel in the form of roads radically transformed the nation. Paved roads have profoundly shaped where and how hundreds of millions of people live and work, what they value, even what they eat—in America, and increasingly, everywhere.

The need for a flat, durable track beneath your wheels is an ancient one. People have been manufacturing hard roads since as far back as 4000 BCE; the streets of the Mesopotamian cities of Ur and Babylon were paved with mud bricks glued together with naturally occurring bitumens—sticky, gooey, tar-like materials also known as asphalt.[13]

The word *pavement* comes from the Romans, who developed the first major road network to connect their empire. Their roads were surfaced with a top layer of stones they called *pavimentum*.[14] Modern paved roads have their origins in eighteenth-century England. An Englishman named John Metcalf developed a system of well-drained roads built with large stones covered by a layer of gravel, which he used to cover 180 miles of Yorkshire byways.

In 1816, a Scotsman, John Loudon McAdam, came up with the

idea of putting down a layer of broken, sharp-edged stones, then running a horse-drawn roller over them to compact them together to form a strong surface. Other road builders improved on the process by adding hot asphalt to keep dust down and to glue the stones together. The method was dubbed *tarmacadam*, after its progenitor. From this evolved the technique of combining asphalt with sand and gravel to make asphalt pavement, aka blacktop, aka bituminous concrete—but usually just called asphalt. Modern asphalt pavement is often more than 90 percent sand and gravel.[15]

Relatively easy and cheap to make, and highly effective, asphalt caught on. France laid down one of the first asphalt roads as part of its Paris–Perpignan highway in 1852,[16] and within a few decades the material was used to pave many of the roads of London and Paris. In the United States, asphalt pavement was introduced in front of the Newark, New Jersey, City Hall in 1870. Washington, DC's Pennsylvania Avenue came soon after. It wasn't long before New York City decided to ditch brick, granite, and wood in favor of asphalt paving on its streets. One advantage asphalt had over wood was that it didn't soak up urine from the endless parade of horses that were the primary form of transport at the time. And unlike brick or stone, asphalt had no gaps between blocks for manure to get stuck in, a serious health hazard.

In those days, almost all asphalt used in the United States was naturally occurring, imported by ship from two giant lakes of it in Trinidad and Venezuela. (Los Angeles's La Brea Tar Pits[17] are another natural lake of bitumens.) As demand grew, the imported material was gradually replaced with man-made asphalt derived from another booming industry: oil. By lucky coincidence, bitumens are created as a by-product of refining gasoline from petroleum. So the more gas that was manufactured to fuel cars, the more asphalt there was available to make roads for them to run on.[18]

Meanwhile, other road builders were experimenting with that material that was getting so much buzz in the construction trades: concrete. An inventor named George Bartholomew installed the world's first concrete street in 1891 in Bellefontaine, Ohio. It was such an untrusted novelty that city officials allowed the concrete to be laid only after Bartholomew agreed to donate all the sand and other materials, and to post a $5,000 bond[19] guaranteeing it would last at least five years. The street is still in place today.

There's been a spirited rivalry between the asphalt and the concrete industries in the road-building market ever since. (Asphalt roads are the black ones; concrete roads are gray.) In the 1950s the concrete industry's main trade organization ran full-page magazine ads featuring movie star Bob Hope declaring, "I don't know how they get new-type concrete so flat and smooth riding, but I like it. Makes driving easy, really relaxing." The ad goes on to brag that "concrete is one of the best friends a taxpayer can have," noting it has 60 percent lower upkeep costs than asphalt.[20] These days, asphalt producers like to boast that 93 percent of all 2.2 million miles of America's paved roads are surfaced with their product.[21] They don't mention that it's often just an overlay on top of concrete base.

Both asphalt and concrete are basically just gravel and sand stuck together. The difference is the binding agent. In concrete, it's cement. In asphalt pavement, it's bitumens.

The basic trade-off is that in general, asphalt is cheaper to lay down and to maintain, and provides a smoother, quieter ride.[22] Concrete, on the other hand, lasts longer and doesn't need as much repairing in the first place. The choice often comes down to how much money a given government agency has handy.

Both types of pavement began creeping over city streets in the late 1800s, but outside of urban areas at that time, there was almost nothing but dirt to travel on. Roads just weren't that important. For most of American history, if you wanted to move lots of people or large quantities of goods any significant distance, you did it via water. Rivers, lakes, canals, and seacoasts carried trade and travelers between settlements. Then along came the railroads in the mid-1800s. Trains connected existing centers and made it easier for people to settle further inland. Sometimes the iron horses supplanted waterways altogether. Roads, such as they were, were for local travel and hauling small loads via horse, wagon, or foot.

That state of affairs, however, just couldn't last in a country where everyone suddenly wanted a car. In 1900, only eight thousand motor vehicles were registered in the United States. But sales boomed as the product improved. Technical advances like replacing hand cranks with electric starters made horseless carriages ever more appealing, especially to women. In 1908, Henry Ford introduced the Model T, a relatively cheap car specifically aimed at getting the masses behind the wheel.[23] That's when the automobile really caught on. By 1912, there were nearly a million cars on American roads—10 percent of them Model T's.[24] They jostled for space with the new trucks that farmers were investing in to haul their produce, and which businesses were turning to as an alternative to railroads. At the time, there were still 21 million horses hauling people and cargo, but it was clear automobiles were becoming ever more important.

Motor vehicles, however, could not get far without more, and stronger, roads. A car without pavement is like a pair of skis without snow. You can get somewhere using it, but not quickly or easily. The ascent and ultimate dominance of the auto required the deployment of vast legions of sand. Sand and gravel in the form of pavement is the crucial ingredient that made motor vehicles useful, the infrastructure that turned them from a specialized amusement for rich eccentrics into an all-purpose conveyance for everyone.

As the automobile grew in popularity, national organizations sprang up to lobby for "good roads." The first concrete highway, a 24-mile-long, 9-foot-wide stretch, was laid down near Pine Bluff, Arkansas, in 1913. By the following year, the country had some 2,348 miles of concrete roadways.[25]

Cars and paved roads fueled each other's growth, symbiotically supporting each other. The more cars people bought, the more paved roads they wanted. The more paved roads that were built, the more people wanted cars. The cycle has continued up to the present day. By now, in many places roads are pretty much the only way to get around, and people have no choice but to use cars.

As late as 1919, though, as Eisenhower learned on his motorized odyssey, you still couldn't count on finding a paved road to take you from state to state, let alone across the country.

It was around the time of Eisenhower's convoy adventure that Carl Graham Fisher decided to take matters into his own hands. Fisher was a man who loved speed, and was thrilled with the fast-moving new machines on wheels that came into vogue at the turn of the twentieth century—first bicycles, then cars. Fisher did as much as anyone in America to popularize these new inventions, and make it possible for them to reach their full potential, by becoming one of America's most important early road builders. He sounded a call to arms that mobilized millions of tons of sand into becoming some of the nation's first highways for his beloved automobiles.

Born in Indiana in 1874, Fisher was the Richard Branson of his day—part forward-looking entrepreneur, part showman-salesman, a high-living capitalist with a daredevil streak and an intuitive knack for making his projects look glamorous. He was fabulously rich and famous in his time, though barely anyone remembers him today.

Fisher dropped out of school at age twelve to put his talents to what he considered better use: making money. By fifteen he was selling newspapers and tobacco on trains. He'd been a daredevil since he was a kid, fond of walking tightropes and sprinting backward at full tilt. The wind-in-your-face, pulse-pounding speed of bicycling—a sport surging in popularity at the time—intoxicated him. Within a couple of years he had saved up enough to open his own business in Indianapolis: a bicycle repair shop.

Fisher made himself into his own best advertisement, grabbing public attention with one crazy stunt after another. "He built a bike so big he had to mount it from a second-floor window, then rode it through the city's streets," writes Earl Swift in his history of US highways, *The Big Roads*. "He announced he'd ride a bike across a tightrope strung between a pair of downtown high-rises and, against all reason, actually did it while a crowd watched, breathless, from twelve stories below. Now a minor celebrity, Fisher put out word that he'd throw a bike off the roof of a downtown building and award a new machine to whoever dragged the wreckage to his shop. This time the police tried to stop him, planting sentries outside the building the morning of the stunt. They were no match for the budding showman; Fisher was already inside and at the appointed hour tossed the bike, then escaped down a back staircase. When the cops showed up at his shop, a telephone call came in. It was Fisher, with word that he was waiting at the precinct house."[26]

Fisher was having a ball, and making a bundle, but like other bikers he was frustrated by the state of the roads. Even in cities, they were often paved with cobbles or brick, surfaces that would set a cyclist's teeth rattling. Bicycling was exploding at the turn of the century, and riders formed a potent lobby. Fisher joined the League of American Wheelmen, one of several organizations pushing for good roads. His interest grew sharper as he started playing around with even newer riding machines. First it was motorcycles, and then, inevitably, automobiles.

Once he'd bought his own three-wheeled, 2.5-horsepower car, Fisher knew these machines were going to be big. In 1900, he shut down his bike shop and replaced it with the Fisher Automobile Company, one of America's first car dealerships.[27]

Fisher and a couple of pals from his bike racing days promoted the machines with appearances at county fairs, where he won bet after bet that his horseless carriage could outrun the fastest horse the locals could find. His dealership did well, but his big break came when he got the chance to invest in a company making the first practical auto headlights. Today it's known as Prestolite, a multinational manufacturer of auto components.

Fisher used his headlight profits to pursue a couple of pet projects. One was building a speedway outside his hometown to host a giant race he organized—the Indianapolis 500. Another, less sexy but more important, was a campaign to build a 3,400-mile coast-to-coast highway from New York's Times Square to San Francisco's Golden Gate Park.[28]

The project, which he grandly titled the Lincoln Highway, was of course too big for one man to take on, no matter how wealthy. Leveraging his newfound prestige and connections, Fisher got backing from politicians including President Woodrow Wilson, celebrities like Thomas Edison, and the heads of major car, tire, and cement companies. In 1913, Fisher himself led a thirty-four-day convoy from Indianapolis to Los Angeles to scout possible routes and drum up publicity. The first piece of the concrete road was built the next year,[29] in northern Illinois.

Though the Lincoln never made it all the way between the coasts, it came close, building new stretches of road and incorporating and improving on existing ones. By the 1920s the Lincoln had become "the nation's premier highway," according to the Federal Highway Administration's official history. It did a lot to convince federal and local governments, and the public, that a transcontinental road was not only possible but desirable.

The Lincoln wasn't Fisher's last foray into road building, however. Just a few years after that highway launched, Fisher built another one. This road stretched from Chicago all the way to another American institution Fisher built from scratch. It was a new resort town called Miami Beach, and it, too, was literally made from sand. We'll meet up again with him there later.

Spurred in part by Fisher's project, the federal government threw its weight into road building. In 1916 it created the Bureau of Public Roads, endowed with $75 million to hand out to states to help build intercity highways.[30] In a stirring speech to a gathering of regional highway builders in 1918, Interior Secretary Franklin Lane compared their efforts to those of Napoleon and Julius Caesar, telling them that they were "engaged in a very farsighted, important bit of statesmanship, work that does not have its only concern as to the farmer of this country or the helping of freight movement during this winter alone, but may have consequences that will extend throughout the centuries."[31]

One of the central difficulties in building those first highways was getting the armies of sand to where they were needed. Each mile of paved road required around 2,000 tons of sand and 3,000 tons of gravel.[32] Hauling all that aggregate out to the rural areas where most of the new highways were being built was no small feat; after all, at the time there were hardly any trucks, and no existing roads on which to transport the aggregate from the mines to the new roadbeds. Builders had to rely on horses and wagons, or build special rail lines to bring trains to the roadbeds. Locomotives would haul in carloads of rock, sand, and cement to be mixed on-site.[33]

Still, with federal money priming the pump, the project moved forward rapidly. The nation's surfaced road mileage nearly doubled from 1914 to 1926, from 257,291 miles to 521,915 miles.[34] Yet the road builders were barely keeping up with the need. The number of automobiles by then had reached nearly 20 million. Even the Depression barely slowed down the automobile. "By 1939, automobile driving had long since passed from an amusing activity for the enjoyment of the indolent and wealthy to become an essential part of American life. Even the Joads in Steinbeck's *Grapes of Wrath* drove to California in their own truck," writes Tom Lewis in another history of American roads, *Divided Highways*.[35]

Roads became a major industry unto themselves. Hundreds of thousands of men worked building them (including chain-ganged prisoners forced to break rocks for roads).[36] More jobs were created in the gas stations, repair shops, restaurants, hotels, and motels that grew up alongside the new highways. Hundreds of other businesses grew fat supplying the raw materials to the road makers—cement, asphalt, gravel, and of course, sand.

You may recognize the name of Henry J. Kaiser, or at least his last name, in those of the gargantuan enterprises he founded— Kaiser Steel, Kaiser Aluminum, the Kaiser Permanente health system, and the Kaiser Family Foundation. Kaiser was one of the twentieth century's most powerful industrial moguls, but he started out literally at ground level, as a supplier of sand and gravel to the road-paving trade.

Born to working-class German immigrants in New York in 1882, Kaiser quit school at thirteen and headed west to seek his fortune. He wound up in Washington State working for a gravel and cement dealer. One of his earliest big projects was building a new sand and gravel mine. Never short on confidence, he struck out on his own, taking over a failed road-building business and reviving it, landing contracts to construct streets in Vancouver and other Canadian cities. But he soon began looking farther south. When the new Bureau of Public Roads began doling out millions for highway building in 1916, Kaiser saw the vast potential of booming California.[37] He relocated to Oakland, and in 1923 scored a contract to build a road through the nearby Livermore Valley. The valley turned out to be rich in easily accessed gravel and sand, so Kaiser simply bought up tracts of farmland, stripped off the topsoil, and mined the aggregate. There was enough to not only build his road but also a business. Thus was born Kaiser Sand and Gravel, supplier to the local construction industry[38] and a foundation block of Kaiser's empire.

During this time, Kaiser also forged an alliance with an inventor named Robert LeTourneau,[39] who developed some of the earliest heavy road-building equipment—giant mobile machines that could move tons of dirt and sand far faster than any work gang with a pack of mules. Those machines helped Kaiser become a major builder and materials supplier in the West. In the late 1930s, he won the job of providing the 11 million tons of sand and gravel needed to build California's Shasta Dam. Kaiser figured it would be simple, since he already owned a sizable aggregate mine near the dam site north of Redding; all he had to do was load it up on trains and pay for the transport. But the local railroad quoted a price Kaiser thought too high. So he came up with an audacious work-around. He built a conveyor belt nearly ten miles long, the longest the world had ever seen,[40] to carry a thousand tons of sand and rock per hour up and down rugged hills and across several creeks to the dam site. Later, Kaiser parlayed his expertise with aggregate into a prize gig as one of the main contractors building the Hoover Dam.

Meanwhile, in Europe, while Germany's politicians—including an ascendant Adolf Hitler—were horrifying the world, their engineers were winning applause for the nation's new autobahn, the first superhighway. The autobahns premiered some of the key features that still define modern freeways. They were one-way roads at least two lanes wide, kept apart from their twins coming the other way by a wide median. Their curves were banked to allow higher speeds. They were separated from regular roads and accessible only from limited on- and off-ramps. And they were surfaced with solid concrete. They were the smoothest, fastest roads ever built.

Americans soon started copying the style, building freeways like the Pennsylvania Turnpike and Los Angeles's Arroyo Seco Parkway. The *Los Angeles Times* ran a front-page story[41] about the 1940 opening of that "impressive boulevard," breathlessly noting that the Rose Queen had untied the red silk ribbon to officially open the six "glass-smooth miles" of "six-lane highways, important to traffic, history and national defense." California's governor Culbert Olson declared that it would whisk motorists from the heart of Los Angeles to central Pasadena in as little as seven minutes "in easy, nerve-free comfort and safety." Nearly eighty years later, the parkway still carries people from downtown LA to downtown Pasadena. The trip takes a whole lot more than seven minutes, though, and not one of them is glass-smooth or nerve-free.

A mong those deeply impressed by the German autobahns was Dwight D. Eisenhower, whose career had come a long way since that cross-country convoy. He had risen to become commander of the Allied forces in World War II, an exalted perch from which he saw how quickly German forces could get around on their well-designed highways, and how much more resilient the road network was compared to rail lines. Trucks can drive around bomb craters, after all, but trains can't get past damaged track. (The Nazis knew sand was important for more than roads, incidentally. During the war, the Germans built specially designed tanks to spread sand on icy roads so that military vehicles could use them.)[42]

Eisenhower was elected president in 1952, and he took those lessons with him to the White House. "After seeing the autobahns of modern Germany . . . I decided as President to put an emphasis on this kind of road building," he later wrote. "The old convoy had started me thinking about good, two-lane highways, but Germany had made me see the wisdom of broader ribbons across the land."[43]

Luckily for Eisenhower, much of the political and administrative groundwork for such a project had already been laid. Thomas Harris MacDonald, the longtime head of the Bureau of Public Roads, had spent years cultivating support for a national highway system, helping to wring billions of dollars out of Congress to support state road-building efforts and coauthoring a major report advocating a nationwide toll-free highway network. Lobbyists from the asphalt, concrete, contracting, automobile, and oil industries gave their support.[44] So did many of the 72 percent of American families that owned cars by the mid-1950s.

Even so, it took a couple of years and several unsuccessful attempts to get Congress to agree to fund the proposed National System of Interstate Highways. Tweaking the routes of the proposed highways so that they ran through carefully chosen cities in every state helped secure many representatives' votes. Others were swayed by the promise of all the construction jobs the project would create. There was also the Cold War argument that the roads were essential for national defense. If the Russians sent nuclear missiles screaming toward American cities, the theory went, big freeways would help millions of civilians evacuate quickly. To make sure Congress got the point, the project was renamed the National System of Interstate and *Defense* Highways.[45]

Congress finally passed the bill to fund interstates in 1956. The act allocated $25 billion to build a highway system stretching 41,000 miles. All of the roads would be limited-access divided highways, with twelve-foot-wide lanes and sight distances to permit speeds of up to 70 miles per hour. The bill also raised taxes on gasoline, diesel fuel, and tires to help pay for the project. Federal planners expected the whole enterprise would be completed by 1972.

Roads engineered to such specifications would require a phenomenal amount of sand and gravel. In addition to all the grains embedded in the eleven inches of concrete on the roads' surface, a further 21 inches of aggregates were needed for the underlying road base. At the outset of the project, the Federal Highway Administration estimated that, all told, the interstate would consume enough sand, gravel, crushed stone, and slag "to make 700 mounds the size of the largest Egyptian pyramids."[46]

Naturally, as construction got under way in earnest, the

demand for sand to pave all those miles soared across the country. Consumption of sand and gravel in the US hit a record high of nearly 700 million tons in 1958, a figure almost twice the 1950 total. By then, according to a federal Bureau of Mines report, so much had already been used that "sources of aggregate were limited in some states" and "nearly depleted in other areas."[47] Entire new types of monster dump trucks, capable of carrying huge loads off-road, were designed to meet the need to move all that aggregate.

At the same time, commercial jet airplanes were coming into everyday use. They required enormous runways, much longer and wider than their predecessors, as well as expanded airports—all of which sand and gravel would be called upon to build. With lucrative contracts for highways and runways being offered across the country, contractors surged into the paving industry. Major corporations decided they wanted a piece of the action, and started buying up sand and gravel companies. Remember Henry Crown? His Material Service Corporation merged with the mammoth defense contractor General Dynamics during this period. (It was sold off in 2006 to global giant Hanson for $300 million.)

The soaring demand for sand also meant major business for the hundreds of smaller local outfits run by men like Ralph Rogers, an eighth-grade dropout who got his start in 1908 crushing rocks by the side of the road near Bloomington, Indiana. His company grew as a supplier of aggregate for military bases, but its big break came when it became one of the first to supply the interstate system in the 1950s. That set what is now the Rogers Group on a trajectory that has made it one of America's largest privately owned aggregate companies, with 1,800 employees and more than a hundred quarries in six states.[48]

Figuring out exactly how to build those roads took some doing.

The Bureau of Public Roads set up a testing center near Chicago where researchers experimented with different types and proportions of sand, gravel, cement, and other ingredients to figure out how much of a beating from heavily loaded trucks each paving mixture could stand up to and for how long. They built a series of looping test tracks composed of various asphalt and concrete mixes, and then set a company of soldiers to drive trucks over them—nineteen hours a day, every day for two years.[49] The bureau used the data to set pavement design standards.[50]

Those standards included specifications for the aggregate acceptable for use on interstates. Like soldiers called to the nation's service, sand grains for the new highways had to meet physical requirements of size and strength. That forced sand and gravel companies to invest in more sophisticated sorting machines. Mining and sorting equipment became increasingly automated, producing ever more aggregate with ever fewer workers.

Official construction of the new highways began in the summer of 1956. At first the program was very popular. But the giant highways cut sometimes painful swaths across America. Land was taken for the roads' rights-of-way, forests were cut down, fields were paved over, neighborhoods were bulldozed. Whole sections of cities, suddenly isolated behind concrete barriers, withered.

Disenchantment grew fast. Social scientist Lewis Mumford, one of the earliest and most prominent critics of the interstates, denounced "those vast spaghetti messes of roads and clover crossings and viaducts that provide excellent material for aerial photography but obliterate the towns they pass through." He hated their impact on major cities, too, calling it "pyramid building with a vengeance: a tomb of concrete roads and ramps covering the dead corpse of a city."[51] Journalists published scathing exposés of graft and wasteful spending during construction. Citizens rose up in what came to be called "The Highway Revolt," fighting back against plans to shove roads through their cities. They won their first victory in San Francisco in 1959, stopping plans to build a double-decker freeway that would have cut off the downtown from the waterfront. Other campaigns thwarted or forced plans to change in New York City, New Orleans, and other cities.[52] Over time the highway builders responded, to a certain extent, introducing measures to reduce noise, minimize environmental damage, and preserve historic areas.[53]

The interstate was finally officially completed in 1991, nearly twenty years behind schedule. It stretched 46,876 miles and cost nearly $130 billion.[54] At the time, it was the biggest public works project in American history. A lattice made of billions of tons of sand and gravel now connected the United States to itself far more intimately than ever before.

The interstates have turned out to be a double-edged sword. It's hard to think of any other project or development that has so profoundly transformed America as have freeways generally and the interstates specifically. The car was and remains the foremost avatar of modernity, and asphalt and concrete are its little-noticed helpmeets. Freeways have altered where we live, work, and shop, and how we get to the places where we do those things.

Much of that has been to the good. Paved roads have enabled goods to reach distant markets, knitted regions together, and made it far easier to visit loved ones and distant places. They have also saved countless lives. One benefit modern freeways don't get enough credit for is the dramatic extent by which they have reduced the number of road deaths. Thanks to their well-engineered banks, wide lanes, gentle curves, separation from automobiles coming the other way, and careful control of merging ones, interstates are far safer than the roads they replaced. In fact, according to the Federal Highway Administration, the interstate is the safest road system in the country, with a fatality rate of 0.8 deaths per 100 million vehicle miles traveled, a rate almost half the national average. Compare that to the rate in 1956, when the interstates were launched, which was 6.05.[55]

(Of course, to get those results you also need things like safety belt laws and traffic lights. Otherwise, highways can quickly become charnel houses. Each year around the world, nearly 1.3 million people die and as many as 50 million more are injured in car crashes. More than 90 percent of those deaths happen in less-developed countries,[56] where traffic lights are rare, seat belts are little used, and the simple act of crossing the street often requires a pulse-pounding sprint through traffic.)

At the same time that freeways have brought these benefits, though, they have also hollowed out cities, killed off countless small towns, wreaked environmental havoc, and spawned a car-dependent culture based on sprawling suburbs and soulless shopping malls.

In the course of the construction of the interstates, urban neighborhoods, especially ones full of African American, Hispanic, and low-income residents, were cut through, paved over, or isolated and left to stagnate. "Planners and residents alike found that new highways . . . could transform a once vibrant neighborhood into a cold, alien landscape," writes Lewis.[57] "White flight" took hold as those who could afford to moved out of cities to commuter suburbs made accessible by the new freeways. The loss of all those affluent residents gutted the tax base of many cities, undermining public schools and other services. Downtown shopping districts emptied out as customers flocked to malls built close to highway off-ramps.

Small towns got hit, too. Those that had grown up alongside railroads or rural routes but were bypassed by the interstates withered. The railroads lost out as well, both in the freight and passenger businesses. Today, trucks carry 70 percent of all US freight, seven times more than trains.[58] By 1986, America's interstates, though they made up only 1 percent of the nation's freeways, carried 20 percent of its truck traffic. Manufacturing jobs also followed the freeways. Companies abandoned cities to build their factories on cheaper land in rural areas easily reached by the new roads.

Roads built of sand opened up whole new tracts of the country for suburban settlement. Buildings made with sand made it possible for people to live in those areas. You no longer needed a nearby source of trees or clay to build with; you just needed an open piece of land and a road that concrete trucks could drive in on. The number of Americans living in suburbs mushroomed from 30 million in 1950 to 120 million in 1990.[59] The numbers have kept climbing ever since.

In many ways, suburbs are great. They provide millions of people with relatively quiet, safe, affordable homes, often endowed with swaths of private outdoor space their tenement-dwelling grandparents could only dream of.

In others, they're terrible. Suburbs devour land and make people dependent on cars, the source of so much pollution and greenhouse gases. The average driver now puts 14,000 miles on his or her car each year—a 40 percent increase just since 1980.[60] That burns up around 172 billion gallons of gasoline per year,[61] almost double the amount in 1970.

Whatever else you can say about suburbs, their low density and dependence on cars make them an especially sand-intensive form of settlement. Think of all the sand that goes into those wide roads and all those low-slung, spread-out houses, each with its own driveway. Every one of those houses contains hundreds of tons of sand and gravel, from its asphalt driveway to its concrete foundation to its stuccoed walls to the grains on its roof shingles.

The open spaces of suburbia also made possible an explosive proliferation of swimming pools, which require large amounts of sand in the form of concrete. (Pools also generally use sand filters to keep the water clean.) In 1957, there were only about 4,000 private swimming pools in the United States. By the next year, the number had shot to 200,000.[62] It's now more than 8 million.[63]

American sand and gravel production grew in step with the spread of suburbs. It had been increasing steadily since the beginning of the century, but after World War II, it abruptly skyrocketed.[64] Today the annual US total hovers around a billion tons, the vast bulk of which is used domestically.

Ironically, while the growth of suburbs meant big business for sand and gravel producers, it also created some significant headaches for them, as their quarries were rapidly surrounded by new housing developments full of people who didn't appreciate all their noise and dust and started lobbying against further mining. In the late 1950s, for the first time, the National Stone, Sand, and Gravel Association set up a public relations team to "meet the challenge threatening the existence of many producers," as the trade magazine *Rock Products*[65] put it.

One unexpected side effect of laying down all those sand and gravel roads across the nation was the proliferation of interchangeable, deliberately monotonous chain stores, fast-food restaurants, and gas stations that sprouted up in self-contained clusters near the interstates' off-ramps. These chains explicitly aimed to provide an experience as predictable, safe, and easily accessed as the highways themselves, those great rivers of pavement that carried customers to their doors. It was no accident that one of the advertising slogans for Holiday Inn, a chain that found success by building hundreds of motels near freeways and interstates, was "Holiday Inn. The best surprise is no surprise."

In this way, freeways have helped to rob many places of their personalities, smothering regional character under a blanket of sand and gravel. The interstates are designed to be monotonous, engineered to the same standards, governed by the same speed limits, with signs in identical colors and fonts indicating the distance to the next city. As a result they induce highway hypnosis, providing an experience less like motoring than like sitting on a vast concrete conveyor belt, cruise-controlling along with no effort required beyond keeping one eye on the road and another on your gas gauge, for mile after mile after mile. That numbing sameness reduces the landscape to a blur interrupted at regular intervals by overbright outposts of gas stations and fast-food chains, replicated in slightly different configurations right across the entire country, so that you can have breakfast at a Denny's in the morning in Nashville and dinner at what appears to be exactly the same Denny's that evening in Minneapolis. The interstates connect towns and cities but are disconnected utterly from them and the land they pass through.

Along with the exit-ramp convenience colonies, highways also fueled the growth of shopping malls. The first enclosed, climate-controlled mall opened in 1947 in Minnesota, and in short order such places became a fixture of American life from coast to coast. Many of them could not exist without the highways that bring them customers from far and wide. More concrete begetting the use of more concrete, more sand begetting the use of more sand.

Today, 2.7 million miles of paved roads crisscross the United States, traversed by 256 million motor vehicles that cumulatively travel nearly 3 *trillion*[66] miles every year. The interstates make up just 1 percent of those roads, but they carry one-quarter of all highway traffic. The United States is not building new highways at nearly the pace of previous decades, but still adds over 30,000 lane-miles of highway per year. Counting the road base as well as the concrete and asphalt on top, each of those lane-miles requires an average of 38,000 tons of aggregate.

The demand for more roads isn't likely to slacken any time soon. Traffic keeps getting worse. According to the Texas A&M Transportation Institute, in 2015, delays due to congestion kept drivers stuck in their cars for nearly 7 billion extra hours, wasting over 3 billion gallons of fuel.[67] That works out to 42 hours annually per commuter, double the figure in 1982.

This car-centered, highway-enabled, sand-intensive way of living is the model much of the rest of the world is now trying to emulate. All those millions of increasingly affluent Vietnamese, Brazilians, Indians, and above all, Chinese want their own cars and the lifestyle with which they're associated.

In almost every country on earth, the number of motor vehicles in use is increasing. There are at least 1.2 billion already on the move, and that number is projected to more than double by 2050. Mexico City is currently adding two cars for every new resident each year; India is adding three.

All those vehicles need pavement, and they're getting it. Between 2000 and 2013, the world added 7.4 million[68] miles of paved roads; that's more than triple the total in the United States. Plans are in the works in Africa to build the first-ever highway stretching from Cape Town, South Africa, to Cairo, Egypt, and another that will wend its way across the Sahara. China is once again in a league of its own. In the last decade alone, China has built 1.3 million miles of paved road, tripling its network. It is now the world's leading asphalt consumer. The Chinese expressway system is now longer than the US interstate system, and in some places makes it look downright puny. There's a stretch of highway linking Beijing with Hong Kong that is a full 50 lanes wide. The International Energy Agency[69] estimates that by 2050, the world will add more than 15 million miles of paved roads. Some 30,000 square miles of new parking spaces—also made with sand and gravel—are also in the pipeline.

The use of sand in the form of concrete and asphalt has completely transformed where we live and work and how we move around. It has given us the power to conquer geography and overcome the elements. Almost at the same time as these changes began to take hold, the use of sand in another form—glass—began to change our lives in equally radical ways.

CHAPTER 4

The Thing That Lets Us
See Everything

Deep below the earth, in a cavernous West Virginia mine one day in 1868, a miner slammed his pick so hard into a tunnel wall that a chunk of coal flew out and smashed into the right eye of one Michael Owens, knocking him unconscious. It was hardly an unusual sort of accident, but nonetheless, Owens's mother was very upset. After all, he was only nine years old.

It took some time for the boy to recover, and once he had, his mother insisted he not return to such a dangerous environment. Which didn't mean going to school, of course. Owens was the third of seven children in a poor immigrant family. His parents had fled Ireland to escape the potato famine and the oppressive British in the early 1840s, settling in West Virginia. It was a tough place to make a living, and it was perfectly common for boys to work in the coal mines alongside their fathers to earn extra money.

The northern part of West Virginia is also rich in another less famous mineral. The loosely consolidated grains of the Oriskany sandstone, a hundred-foot-thick formation laid down more than 300 million years ago, are some of the purest quartz sands in the United States. Miners started digging them out in earnest after the

Civil War,[1] fueling the growth of a glass industry in the town of Wheeling, where the Owens family lived. Like coal mining and many other industries at the time, glassmakers welcomed child laborers. And so it was that Michael Owens went to work in a glass factory.[2]

Truth to tell, it wasn't much of an improvement, safety-wise. Glass is mainly made of melted quartz sand. Melting those durable grains requires tremendous heat, which in Owens's day was provided by coal. Owens's first job at the factory, at age ten, was as a blower's dog, stoking coal into the glory hole of a furnace. Every day, black dust and ash covered his body and filled his lungs. Wearing knee pants held up with suspenders, he worked six days a week, ten hours a day, starting at five A.M. The temperature inside the factory sometimes topped 100 degrees. He was paid thirty cents a day. "The constant facing of the glare of the furnaces, and the red-hot bottle causes injury to the sight," reported a visitor to a glass factory of the time, noted Quentin Skrabec Jr. in his book *Michael Owens and the Glass Industry*. "Minor accidents from burning are numerous."[3] The glass factories employed boys as young as seven. Adult glassblowers[4] screamed at and beat them. A magazine journalist at the time called it "a boy-destroying trade."

But in Owens's case, at least, it was a career path that paid off. The dirt-poor child laborer would grow up to revolutionize the glass industry, and in the process profoundly change American life. His contributions to his chosen trade were many, but the first one of truly historic significance came in the form of something small enough to hold in your hand. It spawned an industry that today rakes in more than $5 billion a year in the United States alone. And almost by accident, it helped to end child labor in the glass industry. All because an immigrant family happened to settle near a deposit of high-quality sand.

Next to concrete, glass is undoubtedly the application of sand that has most profoundly shaped the modern world. Today, glass is so commonplace that most of us never even think about it—but we should, because it's flat-out astonishing.

Glass is in the buildings we work and live in, the windows we peer out of, the lightbulbs we turn on, the vessels we drink from, the televisions we stare at, the watches we glance at, and the cell phones we can't put down. It is an almost magical substance. It can be shaped and molded into almost any form, from twenty-ton slabs to strands thinner than a human hair, from delicate crystal to bulletproof shields. It makes fiber-optic cables and beer bottles, microscope lenses and fiberglass kayaks, the skins of skyscrapers and the teeny camera lenses on your cell phone.

Glass is the thing that lets us see everything. Without it, we'd have no photographs, films, or television, "no understanding of the world of bacteria and viruses, no antibiotics and no revolution in molecular biology from the discovery of DNA," write historians Alan Macfarlane and Gerry Martin in *The Glass Bathyscaphe*. "We might not even be able to prove that the earth goes round the sun." Even our view of our own bodies would be radically different: glass is the ingredient that makes cheap and abundant mirrors possible.

This miraculous compound is mostly just melted sand; silica makes up as much as 70 percent of the volume of typical window glass. But not just any sand will do. A more refined breed of grain is required than the common construction sand used for concrete. Glass sand belongs to a category called industrial, or silica, sand. To make it into this club, the sand typically must be at least 95 percent pure silicon dioxide, and largely free of certain impurities. (The most common impurity in sand is iron, which imparts a green color; that's why sheet glass seen from the side looks green.) The best silica sands also come relatively uniform in size. Grains that are too big won't melt as easily, and ones that are too small will be blown away by air currents in the furnaces.

Befitting their nobler composition, industrial sands are much more expensive than those used for construction. Though America produces ten times more construction sand than industrial sand each year, the US Geological Survey estimates that the total value of the elite industrial grains is actually higher than that of their lower-grade cousin: $8.3 billion per year, versus $7.2 billion.

The sands chosen for glass have a fundamentally different mission than those used in concrete. Construction sand grains retain their form when made into concrete; they are cemented together with countless legions of their fellow grains and their big brothers, gravel pieces, perpetually working together. The grains that become glass, however, are actually transmuted, losing their individual bodies as they are fused together to form a completely different substance.

Getting them to do that, however, is not easy. It takes temperatures topping 1,600 degrees Celsius to melt silica grains. But mixing sand with additives known as flux, such as soda (aka sodium carbonate), lowers that melting point dramatically. Throw in a little calcium, in the form of powdered limestone or seashell fragments, melt it all together, and when the mixture cools, you have basic glass.[5]

Part of what makes glass so adaptable is that the silicon dioxide that forms it is sort of a solid that acts like a liquid. As materials scientist and engineer Mark Miodownik explains in *Stuff Matters: Exploring the Marvelous Materials That Shape Our Man-Made World*, a regular solid, like ice, can be melted into a liquid—water—and then frozen again, and each time its molecules will re-form into a crystal pattern. "With SiO_2 things are different," writes Miodownik. "When this liquid cools down, the SiO_2 molecules find it very difficult to form a crystal again. It's almost as if they can't quite remember how to do it: which molecule goes where, who should be next to whom, appears to be a difficult problem for the SiO_2 molecules. As the liquid gets cooler, the SiO_2 molecules have less and less energy, reducing their ability to move around, which compounds the problem: it gets even harder for them to get to the right position in the crystal structure. The result is a solid material that has the molecular structure of a chaotic liquid: a glass."[6]

No one knows how people first discovered this miraculous bit of alchemy, but we do know it happened a very long time ago. It was probably an accident, when someone made a fire on a beach where the sand contained some kind of flux that lowered its melting point—soda ash left over from burning certain plants, maybe, or seaweed. And it probably happened in more than one place. Glass beads have been found in modern-day Iraq, Syria, and the Caucasus dating back four thousand to five thousand years. Glass was a must-have ornamental item throughout the ancient world, showing up as glazing on pottery, in jewelry, and as small containers. Ancient Egypt in the time of Rameses the Great, around 1250 BCE, had a substantial glassworks that made perfume bottles and decorative items.

Writing some 3,000 years later, Dr. Samuel Johnson mused: "Who when he first saw the sand and ashes by a casual intenseness of heat melted into a metalline form, rugged with excrescences and clouded with impurities, would have imagined that in this shapeless lump lay concealed so many conveniences of life as would, in time, constitute a great part of the happiness of the world. . . . He was facilitating and prolonging the enjoyment of light, enlarging the avenues of science, and conferring the highest and most lasting pleasures; he was enabling the student to contemplate nature, and the beauty to behold herself."[7]

The Romans, as usual, took the technology to the next level. They made great advances in understanding how to use flux, to the point where they were able to manufacture glass in relatively large quantities and export it all over the empire. They figured out that adding manganese oxide helped clarify the glass, which led to a new invention: semitransparent windows.[8] And they refined techniques of glass blowing that produced the most delicate wineglasses the world had yet seen.

Glass caught on like Pokémon. Glasses so clear they let tipplers see the color of their wine came into permanent fashion across Europe. Windows that let in light but kept out rain and cold were a tremendous quality-of-life boost for people living in more northerly, inclement climates (at least those who could afford them). Artisans mastered the process of coloring glass panes and created the beautiful stained-glass windows that still dazzle visitors at the cathedrals of Chartres, York Minster, and many others.[9]

Glassmaking developed into such a profitable art in Venice that in 1291 the city-state's rulers ordered all of the city's glassmakers to move to the island of Murano. There they were treated like aristocrats—but not allowed to leave, lest they take their coveted craft secrets to rival nations. The sand for the Venetians' famous tableware and decorative items was an exceptionally pure type they brought in from the Ticino River, which flows out of the Alps past Milan.[10] Today's Venetian artisans get their sand from France's Fontainebleau region, which is upward of 98 percent pure silica. (Corning, an American company that is one of the world's largest producers of glass and ceramics, also operates the world's largest ophthalmic glass production center in Fontainebleau.)

Around the same time as the establishment of the artisan colony on Murano, the area around Valdelsa, in Tuscany,[11] emerged as another important European glassmaking center. Glassmakers used the abundant local forests for fuel to melt sands from the Arno River and the beaches near Pisa. Unlike their Venetian counterparts, these artisans were free to emigrate, which many of them eventually did, spreading the glass trade across Europe. The Valdelsa region still provides some 15 percent of the world's leaded glass crystal.

In the fifteenth century, Angelo Barovier, one of the Murano artisans and scion of a family of glassmakers, set about handpicking an elite selection of the purest sands he could find. He processed them carefully, and in time developed *crystallo,* the first truly colorless, transparent glass. This turned out to be a historic breakthrough.

Transparent glass not only made for much better windows; it also made possible high-quality lenses, those unassuming little disks that have essentially endowed us with superpowers. The lenses of microscopes and telescopes enable us to see pieces of the universe we didn't even know existed, objects so tiny or so distant that our naked eyes could never perceive them. These innovations underpinned the scientific revolution.

Telescopes and microscopes were preceded by simpler magnifying lenses in the form of eyeglasses, another tremendously important augmenter of human perception. "The invention of spectacles increased the intellectual life of professional workers by fifteen years or more," write Macfarlane and Martin. Eyeglasses likely abetted the surge of knowledge in Europe from the fourteenth century on. "Much of the later work of great writers such as Petrarch would not have been completed without spectacles. The active life of skilled craftsmen, often engaged in very detailed close work, was also almost doubled," Macfarlane and Martin maintain. The ability to read into one's old age became even more important once the printing press came into widespread use from the middle of the fifteenth century.[12]

It's not clear who invented the first telescope. Amid growing demand for eyeglasses, many people around Europe were experimenting with lenses and mirrors by the late 1500s. The first unambiguous record is from 1608, when an anonymous young spectaclemaker from the Dutch town of Middelburg offered an invention to the commander of the Dutch army: a tube containing two glass lenses, "by means of which all things at a very great distance can be seen as if they were nearby."[13] The army brass immediately recognized the device's military potential. Within weeks at least three other Dutch inventors had applied for government patents on telescopes; none were granted, as so many people obviously knew the secret to making them. It's no wonder there was so much optical experimentation in the Netherlands: Holland boasted a sophisticated glass industry, of which Middelburg was an important center, thanks in part to the local abundance of high-quality river sands.[14]

Telescopes—powerful tools for navigators, military commanders, and even artists painting landscapes—spread at an astonishing speed. By 1609 small spyglasses were being sold in shops in France, Germany, England, and Italy. That spring, an Italian scientist named Galileo Galilei heard about them and began making his own. Rapidly improving prototype after prototype, he soon had a device capable of magnifying images twentyfold. Staring up at the night sky through his creation, Galileo was able to perceive truths about the cosmos that changed history. Among many other discoveries, he determined that the earth revolves around the sun, not the other way around—a heretical notion at the time, one that got him placed under house arrest for much of his later life. Sand showed us our real position in the universe—our planet is just one speck among billions.

There is also controversy about who invented the first microscope, but the first crude version is often credited to a Dutch spectacle-maker named Zacharias Janssen, around 1590. Galileo also experimented with using sets of lenses for extreme magnification. Several versions of early microscopes could be found across Europe by the 1620s, but they were not at first used for scientific research. "The role of microscopes was limited mostly to the demonstration of wonders and curiosities of nature, and natural philosophers and the public delighted to see the known world magnified," writes Laura J. Snyder in *Eye of the Beholder,* a history of lenses.[15]

That began to change rapidly in the 1650s, thanks largely to a young apprentice draper in the Dutch city of Delft named Antoni van Leeuwenhoek. Intrigued by the magnifying lenses his profession used to determine the thread counts of fabrics, Leeuwenhoek started experimenting with his own homemade microscopes. Of necessity, Leeuwenhoek—as well as Galileo and other scientists across Europe—became skilled glass grinders. They made their own lenses from plain glass "blanks," which they shaped into lenses by grinding and polishing them with a variety of abrasives, including ordinary sand.[16]

Leeuwenhoek produced hundreds of microscopes and put them to use to discover red blood cells, bacteria, and sperm. He was also the first scientist to seriously investigate the different characteristics of individual sand grains[17]—using lenses made from sand and shaped with sand to examine sand.

Taken together, the introduction of optical instruments in science showed the world that "behind the phenomena we see with the naked eye is an unseen world, and in this invisible world lie the causes of the natural processes we observe," writes Snyder.[18] Lenses showed us "that the world is not—or not only—as it seems to be."

While glass in many forms was spreading across Europe, the Asian powers of Japan and China didn't pay much attention to this new material, though they knew about it. This must rank among one of history's greatest oversights. Among other things, it meant that they had no microscopes, telescopes, or even eyeglasses until Western missionaries introduced them around 1551. That technological gap may help explain why Europe raced so far ahead of Asia in so many scientific fields in the seventeenth and eighteenth centuries.

In the United States, on the other hand, glassmaking was one of the very first industries that the early colonists established. Groups of Dutch, Polish, and Italian glassmakers[19] set up shop in the first permanent British settlement at Jamestown, Virginia, in the early 1600s, making window glass, bowls, and beads for trade with native peoples. In 1739, the industry picked up significantly when a German immigrant named Caspar Wistar opened a glass factory near Salem, New Jersey, attracted by the plentiful trees to fuel furnaces, oyster shells to provide calcium, and abundant supplies of clean, high-purity quartz sand.[20] Wistar's factory produced hand-blown bottles, which were in ever greater demand among beer brewers in the New World. Thomas Jefferson himself had a sideline brewing beer at his Monticello estate in Virginia, where there was such a shortage of glass jugs that he had to order them from as far away as New York. At one point he even experimented with making glass himself.

European glass imports, however, dominated the American market until the mid-1800s, when the Civil War interrupted trade. At around the same time, Americans started finding major domestic sources of high-quality sand. From 1820 to 1880, the number of glass furnaces in the United States grew fivefold, while the number of people working in the industry multiplied twenty-five-fold.[21]

As the Industrial Revolution took hold across America, whole cities and regions grew up around the glass trade, just as they did around steel, coal, and other mushrooming industries. To manufacture glass profitably, glassmakers need easy access to high-quality sand, cheap energy to run the furnaces, and a transportation network to get the product to market. In the 1880s, the city fathers of a small, relatively young town in Ohio called Toledo realized they had all those resources and more. Eager to build up their settlement, they set about wooing glassmakers from the East to relocate. Toledo, they bragged in newspaper advertisements and personal meetings, had cheap land, cheap labor (including employable children as young as eight), natural gas, and a location on Lake Erie that offered access to canals, rivers, and railroads. Just as important, the city lay near a seam of extremely high-quality silica sand, so pure it was sold to glassmakers as far away as Pittsburgh and Wheeling.

The pitch worked. Glass manufacturers poured in. So many of them set up shop in Toledo—around a hundred by the turn of the twentieth century—that it became known as the Glass City. It remained a vital center of the trade for decades. "Toledo glass was used to make the spacesuits of the astronauts who landed on the moon in 1969, and it was used by Admiral Richard E. Byrd in scientific experiments he conducted at the South Pole in the 1930s," notes Barbara Floyd in her history of Toledo, *The Glass City: Toledo and the Industry That Built It*. "It protected America's Declaration of Independence in the National Archives, and it has been used by revolutionaries around the world to convey their beliefs with Molotov cocktails. It has held the punch served at receptions in the White House, and the alcohol in the brown bags of paupers on street corners everywhere. It insulated the Alaskan oil pipeline, and it is used in solar energy panels. It is displayed in some of the finest art museums in the world, and every day it is tossed into garbage pits."[22]

Among the earliest transplants to Toledo was Edward D. Libbey, owner of a glass factory in East Cambridge, Massachusetts. Libbey's business had been prospering, but his unionized workers were demanding higher wages. Moreover, his energy bills kept rising. Those once-vast forests of New England, a resource previously thought inexhaustible, were rapidly disappearing into industrial furnaces. So in 1888, just like an offshoring corporation today, Libbey moved his operation to a place where costs were lower. It was a fateful move for Libbey, for Toledo, and in fact for all of us.

Seeking skilled workers to staff his new factory, Libbey made a personal recruiting trip to the glass industry hub of Wheeling, West Virginia. He quickly signed up a full roster. He was just getting ready to leave his hotel room when Mike Owens, the former child coal miner, barged in. By then a beefy, square-faced, broad-nosed man of thirty, Owens announced that he was coming to Toledo to work for Libbey. This, at least, is the version Skrabec recounts in his somewhat hagiographic biography of Owens. "Libbey apologized, explaining that he had all the men he needed," writes Skrabec. "Owens replied: 'Oh, no, you don't! You need me!' and . . . the man's appearance and self-confidence just stopped him."[23]

However the job interview actually went, Owens was hired. Hard-driving, ambitious, and extremely intelligent despite his near-total lack of schooling, he quickly worked his way up the ranks to become Libbey's top lieutenant. As a manager, Owens was punctilious and demanding. He had a sunny smile and could be charming, but he also had a serious temper. He wasn't averse to cussing out or literally kicking the ass of a malingering worker.

When Owens started at the Libbey Glass plant in 1889, the place was still making bottles much the same way it had been done in the West Virginia factory where he had started out as a boy—which wasn't much different from the method used back in Jamestown.[24] First, the mix of sand, soda ash, and other ingredients was placed in giant pots inside a furnace, where over the course of many hours it melted into a thick, taffy-like goop. Under the supervision of a master glassblower, or gaffer, a gatherer would stick a six-foot iron blowpipe into a pot, swirl up a glob of this infernally hot molten glass, then roll it into a ball on a metal table.

The gaffer, the most highly skilled member of the crew, then took the pipe and blew the mass into the desired shape, sometimes with the help of a cast-iron mold clamped around the molten glass. The glass might cool down during the blowing, requiring a stick boy to put it back into the furnace to soften it up again. Once the glass was in the right basic shape, the gaffer and his assistants would refine it with wooden tools, reheating as necessary. A carry-in boy would then take the still-hot finished piece to another furnace, where it would be gradually cooled and hardened, a process called annealing. A standard crew of five to eight men and boys working ten-hour shifts could produce about 3,600 bottles a day—about one per minute. Not exactly an efficient way to mass-manufacture a consumer product.

Owens figured he could do better. Automation was replacing human hands everywhere, increasing production at an explosive rate in industries of all kinds. Owens was no engineer and had only a rudimentary grasp of the chemistry of glass.[25] But he had worked every stage of the glassmaking process and understood it viscerally.[26] With Libbey's support and the resources of a now-sizable company to draw on, he set to work on making a bottle-making machine.

It took five years and $500,000—a colossal sum in those days—but in 1903 the first Owens Bottle Machine was ready. It sported six rotating arms, each fitted with a mold and a pipe. Owens's key breakthrough was figuring out a way for the machine to gather up the molten glass, something that had stumped other would-be bottle automators. He installed a small pump on each arm; pulling the plunger back created a vacuum that sucked the glass up into a mold, then pushing it down sent a burst of air in to blow the glass into the right shape.[27] Instant bottle. The machine then cut the bottle loose and put it on a conveyor belt leading to the annealing furnace.

The very first model cranked out bottles six times faster than a human crew. By the time Owens had an updated model ready to sell to other bottle makers, the machine could produce a dozen bottles per minute. Not only was the process far faster, it required far fewer workers, especially expensive, skilled ones. It cut the cost of producing a gross (or a dozen dozen) of bottles from $1.80 to 12 cents.

The machine was a smash hit. An industry magazine frothed: "The Owens machine stands alone in a class unapproached by other inventors. It . . . eliminates all skill and labor, and reduces the cost of production practically to the cost of materials used. Not only that, but it puts the same amount of glass into every bottle, makes every bottle of the same exact length, finish, weight, shape and capacity. It wastes no glass, uses no pipes, snaps, finishing tools, glory-holes, gatherer, blower, mold boy, snap boy, or finisher, and still makes better bottles, more of them, at a lower cost, than is possible by any other known process."[28] The invention was such a success that Libbey and Owens cofounded a new enterprise, the Owens Bottle Company, to manufacture bottles and license the technology to other companies. Eighty years later, the American

Society of Mechanical Engineers dubbed Owens's machine an engineering landmark, and declared "Mike Owens's invention of the automatic bottle-making machine in 1903 was the most significant advance in glass production in over 2,000 years."

Suddenly, thanks to Owens's machine, far more bottles than ever before were being made. That meant more glass was needed. And to make that glass, unprecedented quantities of silica sand were drafted into service. In the single year following the introduction of the bottle-making machine, silica sand production in the United States leapt from 1.1 million tons to 4.4 million tons.[29]

Clawing all those grains from the earth wreaked considerable damage on the environment. Starting in 1890, sand miners completely dismantled the Hoosier Slide, a 200-foot-tall Indiana dune near Michigan City that was once a tourist attraction, hauling its grains away in wheelbarrows to sell to glassmakers like the Ball Corporation, makers of the famous Ball mason jar.[30] Like Libbey, the Ball brothers had been coaxed into leaving New York for the Midwest by the cheap gas, high-quality sand, and generous financial incentives offered by local governments. They made millions of jars and other containers with Hoosier Slide sand, which gave the glass a blue tint. Those jars are now prized collector's items. They went out of production after the 1930s because by then the dune was gone. Other dunes along the Lake Michigan shoreline, some as high as 300 feet, were also mined out of existence until public outcry forced the state government to protect them in the 1970s and 1980s.[31]

Elsewhere in Indiana, the *Gary Evening Post* complained in 1913 that "sand sucker" boats were "stealing the bottom" of Lake Michigan to sell to glassmakers.[32] At the time, no permit or payment was required; anyone was free to dredge as much sand as they liked. (Indiana sand also provided fill for the site of the 1893

Chicago World's Fair, and to reclaim the land on which Chicago's famous Lincoln Park was built.)

Owens and Libbey assured their own supply of the crucial resource by creating the Toledo-Owens Glass Sand Company and buying up a mine in the aptly named nearby town of Silica, Ohio. A trade magazine declared the sand quarried there to be "pure white in color and of exceptional quality."[33]

Today, bottles seem a mundane, disposable product. But Owens's machine had consequences so far-reaching it's hard to fathom. It made many people rich, most of whom had nothing to do with bottle-making. It transformed bottles from a luxury to a commodity, altering forever the patterns of what we drink and how, when, and where we drink it.

Within just a few years of its introduction, Owens's machine was making bottles for everyone from milk producers to H. J. Heinz. By 1911, 103 of the machines were at work in the United States and at least nine European countries as well as Japan, cranking out hundreds of millions of bottles annually.

The first impact of the arrival of all these cheap mass-manufactured bottles was, of course, on the jobs of glassworkers. Recognizing the threat to their jobs, just as bricklayers had earlier with concrete, the bottle blowers' unions fought to keep Owens's machine out of their factories. A few of the machines were even sabotaged. But it was a losing battle. By 1917, the number of relatively well-paid skilled glassblowers was cut in half. On the other hand, the market grew so much that the bottle-making industry soon employed more total workers than ever. For the first time, some of those workers were women, who were given jobs sorting and packing the torrent of products flooding out of the factories.

Owens's machine quickly and completely wiped out jobs for another class of workers: children. The unions suddenly became crusaders for eliminating child labor—partly because their low pay dragged down wages for everyone, at a time when workingmen's livelihoods were already in jeopardy. But more important, kids simply were no longer needed in the factories. The dangerous, repetitive tasks that had been given to children were now better handled by machines. In 1880, nearly one-quarter of all glass industry workers were children; by 1919, fewer than 2 percent were.

Owens was celebrated as a crusading reformer. In 1913 the National Child Labor Committee declared that his machine had done more to eliminate child labor in the United States than the organization had through the legislature. The US Bureau of Labor Statistics declared in 1927 that child labor in the glass industry had become "almost a thing of the past, and credit for this is due in no small measure to Michael J. Owens." The irony of all this was that Owens himself didn't see much wrong with child labor. He always insisted his own early career was a fine one for any stouthearted lad. In a 1922 magazine interview, he expounded: "One of the greatest evils of modern life is the growing habit of regarding work as an affliction. When I was a youngster I wanted to work. . . . A great deal of the trouble to day is with the mothers. Too many boys are being brought up by sentimental women. The first fifteen or twenty years of their lives are spent in playing. . . . When they finally start to work, they are so useless and so helpless that it is positively pathetic. The young man who has begun to work when he was a boy has them handicapped. . . . The hard work I did as a boy never injured me."[34] He added: "I went through all the jobs the boys performed, and I enjoyed every bit of the experience."

Child labor hasn't disappeared in all sand-related industries. Today, adolescents toil in sand mines in Morocco, Ghana, Nigeria, India, and Uganda, while miners in Kenya reportedly recruit kids to drop out of school and come to work harvesting sand instead.

No surprise that the introduction of a bottle-making machine would have a major impact on the lives of those in the bottle-making industry. But the impact of Owens's machine was far broader reaching. By making it easy and cheap to convert vast quantities of silica sand into huge numbers of glass containers, it turbocharged many other industries, which in turn transformed what and how much Americans eat and drink.

Before 1900, beer and whiskey were distributed in kegs to taverns; if you wanted some to take home, you had to supply your own jug. Milk was stored in metal cans delivered by milk wagons; it was served in pitchers. There was no such thing as a baby bottle.

Glass is a near-perfect material for packaging food and beverages. It is nonporous and impermeable, and almost nothing reacts with it chemically, which means a bottle will not interact with whatever is inside it. It won't rust or leach BPAs or impart a plasticky taste; the liquid inside will retain its aroma and flavor for a very long time. So the sudden availability of cheap high-quality bottles was a colossal gift to makers of soft drinks, beer, medicines, and other bottled consumables. It wasn't only that the bottles were cheap; they could also be made of uniform size, which made it possible to fill them with machines (some of which Owens also helped design), further bringing down the price of the final product. Ketchup, peanut butter, and all kinds of other foods packaged in glass jars became affordable staples.

Once again, the use of sand in one form led to more of its use in another. Owens's mass-manufactured bottles hit the market at the same time that automobiles were taking over the country and paved roads were spreading. Both developments made it easier than ever to distribute products like bottled drinks far and wide. Trucks loaded with products packaged in sand rolled smoothly from shop to shop on roads made of sand.

The result was an enormous surge in the market for bottled drinks. Sales of a new beverage called Coca-Cola, for instance, went from 300 million in 1903—before Owens's machine hit the market—to 2 billion in 1910. The Coca-Cola Company's official website credits that in part to "major progress in bottling technology, which improved efficiency and product quality."[35]

The beer business also evolved. In the early 1900s, taking home beer generally involved a trip to the local tavern equipped with a jug, bucket, or whatever container was handy. The lack of fancy packaging gave beer a bit of a low-class reputation—it wasn't something the well-brought-up would drink at the dinner table. In the 1930s, brewers began a concerted effort to upscale their products' image by selling it in bottles. The key, of course, was hooking the housewife. "She must be educated to a more easy use of the word, beer, just as she has been educated to the easy use of the word, cigarette," suggested an article from a trade publication in the mid-1930s. "The beer bottle and label are equally important. If the bottle is clear and clean and the label attractive, the housewife will enjoy placing the bottles upon a tray for serving in the home."[36]

Owens's and Libbey's operations came to dominate the manufacture of all types of glass for decades. They followed up the bottle machine with another major project: a machine to automate the making of flat glass, which up to that time had been made by hand. By 1916 they had a good enough model to launch a new company selling sheet glass. Its impact was as profound as the bottle machine, turning windows for houses and cars, as well as glass tableware, from luxury items into everyday basics.

Glass came into even wider use after 1952 when Alastair Pilkington, a British engineer and businessman, developed a technique of pouring molten glass onto a shallow pool of molten tin, resulting in sheets that could be larger and more uniform in size than ever, ideal for big windows in large-scale construction projects. Float plants using this method soon became the industry standard.

Architects quickly began using the newly abundant glass in buildings. Glass-skinned skyscrapers took over city skylines. Plate glass production worldwide mushroomed twenty-five-fold between 1980 and 2010.[37] Today, more than 11 billion square yards of flat glass are consumed every year[38]—more than enough to glaze over the entire city of Houston six times.

The technology of making glass has continued to race forward, and glass is used to do more and more astounding things. Modern life wouldn't be recognizable without some of the advanced applications to which glass has been turned. Owens-Illinois employees in the 1930s developed a threadlike form of glass that is flexible, strong, lightweight, waterproof, and heat resistant, which they dubbed Fiberglas. (Yes, with one *s*. Later, other companies brought their own versions to market and the stuff became known generically as fiberglass.) Others had spun glass into threads before, but the new process allowed for the creation of strands as thin as four microns around and thousands of feet long. As is true of all glass products, it owes its existence to sand. To make fiberglass, silica is melted down along with other substances—boron, calcium oxide, magnesia—to make it more workable and give it other properties desired for specific products, such as greater tensile strength. This molten glass is extruded through a metal sleeve set with tiny holes, and the streams are caught on high-speed winders that spin them into filaments. Once cooled and coated with chemical resin, these strands can be used in all kinds of ways.

Fiberglass-reinforced plastic, tremendously strong but lighter, more malleable, and more weather-resistant than steel, allowed designers to create fanciful new shapes for boats and automobiles. Chevrolet used it in 1953 to produce a sleek, curvaceous sports car called the Corvette. Today it is used for everything from pipe insulation to kayaks. Highly efficient insulation made with fiberglass also helped make possible the movement of millions of people into America's South and Southwest, areas too unpleasantly hot in summer for most folks to consider without a reliable way to keep the heat out. Sand in the form of fiberglass made it easier for people to move to the sand-strewn deserts of Arizona and Nevada.

In 1940 the Owens-Libbey corporate family introduced another major innovation in insulation: double-paned glass called Thermopane. Suburban homes everywhere were soon (and still are) outfitted with huge picture windows and sliding glass doors made of this material.

Owens-Illinois has redoubtable competition in the historic invention department. The name Corning calls to mind ceramic baking dishes. (Ceramics, incidentally, are also largely composed of sand; ground silica provides the skeleton to which the clay and other additives are attached.) Less well known is that Corning is a venerable, pioneering company that not only makes CorningWare and the ubiquitous line of Pyrex bakeware and storage containers, but also some of the most revolutionary glass products in history. Corning was first to market with mass-manufactured lightbulbs and TV picture tubes. Corning also made the heat-resistant windows on NASA spacecraft from the moon rockets to the space shuttle. In 1970, Corning scientists, using high-purity silica, developed the first optical fibers, a breakthrough material capable of carrying enormous amounts of data; much of the Internet's traffic is now carried along fiber-optic cables.

Chances are excellent there's a Corning product in your pocket right now. It's the company's famous Gorilla Glass that makes the screens of iPhones and other smartphones so strong and scratch-resistant. At the dawn of the twenty-first century, sand isn't only all around us. It's *with* us, in our pockets and purses, a key component of the mobile phones that are the symbol and pillar of the digital age.

Even more advanced types of glass are coming. Corning is working on a bendable version of Gorilla Glass, which could be used to make computer tablets that could be folded or rolled up. NSG Group, a Japanese conglomerate, sells self-cleaning glass—windows coated with microscopic amounts of titanium dioxide that react with sunlight to break down dirt. Scientists at England's University of Southampton are working on using nanostructure inside tiny glass disks to store stupefying amounts of digital information—music, movies, whatever—in a far more stable form than even the best hard drives available today.

By now glass is a taken-for-granted amenity in houses and businesses around the world. Most of us spend most of our time indoors these days, but thanks to glass, our offices, factories, and homes have far more natural light, more stable temperatures, and way better views than those of our grandparents.

Glass also offers little life assists in the form of thousands of specialty products, including all those accoutrements of modern middle-class life we barely notice any more—shower doors, picture frames, salt shakers, patio tabletops, mirrors, and on and on.

Michael Owens, the man who did more than anyone to make glass a part of our daily lives, died a rich man, with forty-nine patents to his name, on his way out of a company board meeting in 1923. The Owens Bottle Company, now known as Owens-Illinois, Inc., headquartered at One Michael Owens Way in Perrysburg, Ohio, just south of Toledo, is still the world's leading maker of bottles for alcoholic beverages. It boasts eighty plants in twenty-three countries and more than $6 billion in annual sales.[39]

But glass has long since lost its premier position as the world's beverage container material of choice; plastic bottles and metal cans now make up 80 percent of the market. Glass manufacturing, meanwhile, has largely shifted overseas, leaving Toledo to decline like so many other midwestern industrial towns. There is one silver lining for Toledo residents, though: fewer glass plants mean less air pollution. The blazing furnaces required to melt sand into glass emit substantial amounts of carbon dioxide. They also spew out other compounds, like sulfur dioxide and nitrogen oxides that aren't greenhouse gases, but can form smog, as well as particulates that can damage human lungs. The glass that comes out of the factories may be clear, but the air around them sure isn't.

The industry's center of gravity today is China, which is now both the world's largest producer and consumer of glass, churning out and gobbling up more than half of all the world's flat glass. It so thoroughly dominates glass manufacture today that even the elaborate panels making up the walls of the Glass Pavilion in the Toledo Museum of Art were imported from China in 2006. Back when the twin towers of New York's World Trade Center were built in the 1970s, American glass was used for every inch. Today the lower floors of its replacement, One World Trade Center, are swathed in Chinese glass.

The booming cities of the developing world don't need sand only for concrete; they need it for glass. All those new buildings need windows. The new cars on the new highways need windshields. The new middle classes need tableware, bottles, and cell phone screens. Demand for glass is surging. In 2003, China consumed $1.9 billion worth of flat glass, according to Freedonia[40];

ten years later, the number was nearly $22 billion. The silica sand that makes it has itself become a multibillion-dollar business.

In the twentieth century, concrete, asphalt, and glass utterly transformed the built environment for countless millions in the Western world. Armies of sand brought us skyscrapers and suburbs, windows and bottles for everyone, and the paved roads that the automobile depends on. In the twenty-first century, that sand-based way of life is spreading with blinding speed across the entire world.

In this new era, the sand armies are taking on even more world-changing missions. Sand is now being used to build entire new lands, to pull oil from previously inaccessible pockets of the earth, and to create the digital devices that permeate our lives. A century and a half ago, sand was a useful accessory, a handy tool for a handful of purposes. Today our civilization depends on it.

PART II

How Sand Is Building the Twenty-First Century's Globalized, Digital World

And every one that heareth these sayings
of mine, and doeth them not, shall be
likened unto a foolish man, which built
his house upon the sand.

—MATTHEW 7:26

CHAPTER 5

High Tech, High Purity

Fresh from church on a cool, overcast Sunday morning in Spruce Pine, North Carolina, Alex Glover slid onto the plastic bench of a McDonald's booth. He rummaged through his knapsack, then pulled out a plastic sandwich bag full of white powder. "I hope we don't get arrested," he said. "Someone might get the wrong idea."

Glover is a recently retired geologist who has spent decades hunting for valuable minerals in the hillsides and hollows of the Appalachian Mountains that surround this tiny town. He is a small, rounded man with little oval glasses, a neat white mustache, and matching hair clamped under a Jeep baseball cap. He speaks with a medium-strength drawl that emphasizes the first syllable and stretches some vowels, such that we're drinking *CAWWfee* as he explains why this remote area is so tremendously important to the rest of the world.

Spruce Pine is not a wealthy place. Its downtown consists of a somnambulant train station across the street from a couple of blocks of two-story brick buildings, including a long-closed movie theater and several empty storefronts.

The wooded mountains surrounding it, though, are rich in all kinds of desirable rocks, some valued for their industrial uses, some for their pure prettiness. But it's the mineral in Glover's bag—snowy white grains, soft as powdered sugar—that is by far the most important these days. It's our old friend quartz, but not just any quartz. Spruce Pine, it turns out, is the source of the purest natural quartz ever found on Earth. This ultra-elite corps of silicon dioxide particles plays a key role in manufacturing the silicon used to make computer chips. In fact, there's an excellent chance the chip that makes your laptop or cell phone work was made using quartz from this obscure Appalachian backwater. "It's a billion-dollar industry here," said Glover with a hooting laugh. "Can't tell by driving through here. You'd never know it."

In the twenty-first century, sand has become more important than ever, and in more ways than ever. This is the digital age, in which the jobs we work at, the entertainment we divert ourselves with, and the ways we communicate with one another are increasingly defined by the Internet, and the computers, tablets, and cell phones that connect us to it. None of this would be possible were it not for sand. High-purity silicon dioxide particles are the essential raw materials from which we make computer chips, fiber-optic cables, and other high-tech hardware—the physical components on which the virtual world runs. The quantity of quartz used for these products is minuscule compared to the mountains of it used for concrete or land reclamation. But its impact is immeasurable.

Spruce Pine's mineralogical wealth is thanks to the area's unique geologic history. About 380 million years ago the area was located south of the equator. Plate tectonics pushed the African continent toward eastern America, forcing the heavier oceanic crust—the geologic layer beneath the ocean's water—down underneath the lighter North American continent. The friction of that colossal grind generated heat topping 2,000 degrees Fahrenheit, melting the rock that lay between 9 and 15 miles below the surface. The pressure on that molten rock forced huge amounts of it into cracks and fissures of the surrounding host rock, where it formed deposits of what are known as pegmatites.

It took some 100 million years for the deeply buried molten rock to cool down and crystallize. Thanks to the depth at which it was buried and to the lack of water where all this was happening, the pegmatites formed almost without impurities. Generally speaking, the pegmatites are about 65 percent feldspar, 25 percent quartz, 8 percent mica, and the rest traces of other minerals. Meanwhile, over the course of some 300 million years, the plate under the Appalachian Mountains shifted upward. Weather eroded the exposed rock, until the hard formations of pegmatites were left near the surface.[1]

Long before Christopher Columbus sailed from Spain, Native Americans mined the shiny, glittering mica and used it for grave decorations and as currency. The first European visitor to the area was a Spanish explorer in 1567, but he didn't find much to interest him. American settlers began trickling into the mountains in the 1800s, scratching out a living as farmers. A few prospectors tried their hands at the mica business, but were stymied by the steep mountain geography. "They couldn't find a way to get their stuff to market," said David Biddix, a scruffy-haired amateur historian who has written three books about Mitchell County, where Spruce Pine sits. Biddix's family has lived there since 1802. "There were no rivers, no roads, no trains. They had to haul the stuff out on horseback," he said.

The region's prospects started to improve in 1903 when the South and Western Railroad company, in the course of building a line from Kentucky to South Carolina, carved a track up into the mountains,[2] a serpentine marvel that loops back and forth for twenty miles to ascend just 1,000 feet. Once this artery to the outside world was finally opened, mining started to pick up. Locals and wildcatters dug hundreds of shafts and open pits in the mountains of what became known as the Spruce Pine Mining District, a swath of land twenty-five miles by ten miles that sprawls over three counties.

At a cluttered desk in the living room of his modest house, which sits in a subdivision built on land reclaimed from a defunct mine, Biddix showed me old black-and-white photos he's collected of the wildcat mines of the era—rough-hewn pits scores of feet deep, worked by grim-faced men in overalls wielding shovels and picks. Biddix's grandfather was one of them. His grandmother worked in a mica sheeting house, pulling apart the rocks' translucent, flat, page-like sheets. Mica used to be prized for wood- and coal-burning stove windows and for electrical insulation in vacuum tube electronics. It's now used mostly as a specialty additive in cosmetics and things like caulks, sealants, and drywall joint compound. The sheeting houses are still open, but these days they import the mica from India, said Biddix.

During World War II, demand for mica and feldspar, which are found in tremendous abundance in the area's pegmatites, boomed. Prosperity came to Spruce Pine. The town quadrupled in size in the 1940s. At its peak, Spruce Pine boasted three movie theaters, two pool halls, a bowling alley, and plenty of restaurants.[3] Three passenger trains came through every day.

Toward the end of the decade, the Tennessee Valley Authority sent a team of scientists to Spruce Pine tasked with further developing the area's mineral resources. They focused on the money-makers, mica and feldspar.

The problem was separating those minerals from the other ones. A typical chunk of Spruce Pine pegmatite looks like a piece of strange but enticing hard candy: mostly milky white or pink feldspar, inset with shiny mica, studded with clear or smoky quartz and flecked here and there with bits of deep red garnet and other-colored minerals. For years, locals would simply dig up the pegmatites and crush them with hand tools or crude machines, separating out the feldspar and mica by hand. The quartz that was left over was considered junk, at best fit to be used as construction sand, more likely thrown out with the other tailings.

Working with researchers at North Carolina State University's Minerals Research Laboratory in nearby Asheville, the TVA scientists developed a much faster and more efficient method to separate out minerals, called froth flotation. "It revolutionized the industry," said Glover. "It made it evolve from a mom-and-pop individual industry to a mega-multinational corporation industry."

Froth flotation involves running the rock through mechanical crushers until it's broken down into a heap of mixed-mineral granules.[4] You dump that mix in a tank, add water to turn it into a milky slurry, and stir well. Next, add reagents—chemicals that bind to the mica grains and make them hydrophobic, meaning they don't want to touch water. Now pipe a column of air bubbles through the slurry. Terrified of the water surrounding them, the mica grains will frantically grab hold of the air bubbles and be carried up to the top of the tank, forming a froth on the water's surface. A paddle wheel skims off the froth and shunts it into another tank, where the water is drained out. Voilà: mica.

The remaining feldspar, quartz, and iron are drained from the bottom of the tank and funneled through a series of troughs into the next tank, where a similar process is performed to float out the iron. Repeat, more or less, to remove the feldspar.

It was the feldspar, which is used in glassmaking, that first attracted engineers from the Corning Glass Company to the area. At the time, the leftover quartz grains were still seen as just unwanted by-products. But the Corning engineers, always on the lookout for recruits to put to work in the glass factories, noticed the quartz's purity and started buying it as well, hauling it north by rail to Corning's facility in Ithaca, New York, where it was turned into everything from windows to bottles.[5]

One of Spruce Pine quartz's greatest achievements in the glass world came in the 1930s, when Corning won a contract to manufacture the mirror for what was to be the world's biggest telescope, ordered by the Palomar Observatory in Southern California. Making the 200-inch, twenty-ton mirror involved melting mountains of quartz in a giant furnace heated to 2,700 degrees Fahrenheit, writes David O. Woodbury in *The Glass Giant of Palomar*.[6] Once the furnace was hot enough, "three crews of men, working day and night around the clock, began ramming in the sand and chemicals through a door at one end. So slowly did the ingredients melt that only four tons a day could be added. Little by little the fiery pool spread over the bottom of the furnace and rose gradually to an incandescent lake fifty feet long and fifteen wide." The telescope was installed in the observatory in 1947. Its unprecedented power led to important discoveries about the composition of stars and the size of the universe itself. It is still in use today.

Significant as that telescope was, Spruce Pine quartz was soon to take on a far more important role as the digital age began to dawn.

In the mid-1950s, thousands of miles from North Carolina, a group of engineers in California began working on an invention that would become the foundation of the computer industry. William Shockley, a pathbreaking engineer at Bell Labs who had helped invent the transistor, had left to set up his own company in Mountain View, California, a sleepy town about an hour south of San Francisco, near where he had grown up. Stanford University was nearby, and General Electric and IBM had facilities in the area, as well as a new company called Hewlett-Packard. But the area known at the time as the Santa Clara Valley was still mostly filled with apricot, pear, and plum orchards. It would soon become much better known by a new nickname: Silicon Valley.

At the time, the transistor market was heating up fast. Texas Instruments, Motorola, and other companies were all competing to come up with smaller, more efficient transistors to use in, among other products, computers. The first American computer, dubbed ENIAC, was developed by the army during World War II; it was a hundred feet long and ten feet high, and it ran on 18,000 vacuum tubes. Transistors, which are tiny electronic switches that control the flow of electricity, offered a way to replace those tubes and make these new machines even more powerful while shrinking their tumid footprint. Semiconductors—a small class of elements, including germanium and silicon, which conduct electricity at certain temperatures while blocking it at others—looked like promising materials for making those transistors.

At Shockley's start-up, a flock of young PhDs began each morning by firing up kilns to thousands of degrees and melting down germanium and silicon. Tom Wolfe once described the scene in *Esquire* magazine: "They wore white lab coats, goggles, and work gloves. When they opened the kiln doors weird streaks of orange and white light went across their faces . . . they lowered a small mechanical column into the goo so that crystals formed on the bottom of the column, and they pulled the crystal out and tried to get a grip on it with tweezers, and put it under microscopes and cut it with diamond cutters, among other things, into minute slices, wafers, chips; there were no names in electronics for these tiny forms."

Shockley became convinced that silicon was the more promising material and shifted his focus accordingly. "Since he already had the first and most famous semiconductor research and manufacturing company, everyone who had been working with germanium stopped and switched to silicon," writes Joel Shurkin in his biography of Shockley, *Broken Genius*.[7] "Indeed, without his decision, we would speak of Germanium Valley."

Shockley was a genius, but by all accounts he was also a lousy boss. Within a couple of years, several of his most talented engineers had jumped ship to start their own company, which they dubbed Fairchild Semiconductor. One of them was Robert Noyce, a laid-back but brilliant engineer, only in his mid-twenties but already famous for his expertise with transistors.

The breakthrough came in 1959, when Noyce and his colleagues figured out a way to cram several transistors onto a single fingernail-sized sliver of high-purity silicon. At almost the same time, Texas Instruments developed a similar gadget made from germanium. Noyce's, though, was more efficient, and it soon dominated the market. NASA selected Fairchild's microchip for use in the space program, and sales soon shot from almost nothing to $130 million per year. In 1968, Noyce left to found his own company. He called it Intel, and it soon dominated the nascent industry of programmable computer chips.

Intel's first commercial chip, released in 1971, contained 2,250 transistors. Today's computer chips are often packed with transistors numbering in the billions. Those tiny electronic squares and rectangles are the brains that run our computers, the Internet, and the entire digital world. Google, Amazon, Apple, Microsoft, the computer systems that underpin the work of everything from the

Pentagon to your local bank—all of this and much more is based on sand, remade as silicon chips.

Making those chips is a fiendishly complicated process. They require essentially pure silicon. The slightest impurity can throw their whole tiny systems out of whack.

Finding silicon is easy. It's one of the most abundant elements on Earth. It shows up practically everywhere bound together with oxygen to form SiO_2, aka quartz. The problem is that it never occurs naturally in pure, elemental form.[8] Separating out the silicon takes considerable doing.

Step one is to take high-purity silica sand, the kind used for glass.[9] (Lump quartz is also sometimes used.) That quartz is then blasted in a powerful electric furnace, creating a chemical reaction that separates out much of the oxygen. That leaves you with what is called silicon metal, which is about 99 percent pure silicon. But that's not nearly good enough for high-tech uses. Silicon for solar panels has to be 99.999999 percent pure—six 9s after the decimal. Computer chips are even more demanding. Their silicon needs to be 99.99999999999 percent pure—eleven 9s. "We are talking of one lonely atom of something that is not silicon among billions of silicon companions," writes geologist Michael Welland in *Sand: The Never-Ending Story.*

Getting there requires treating the silicon metal with a series of complex chemical processes. The first round of these converts the silicon metal into two compounds. One is silicon tetrachloride, which is the primary ingredient used to make the glass cores of optical fibers. The other is trichlorosilane, which is treated further to become polysilicon, an extremely pure form of silicon that will go on to become the key ingredient in solar cells and computer chips.

Each of these steps might be carried out by more than one

company, and the price of the material rises sharply at each step. That first-step, 99 percent pure silicon metal goes for about $1 a pound[10]; polysilicon can cost ten times as much.[11]

The next step is to melt down the polysilicon. But you can't just throw this exquisitely refined material in a cook pot. If the molten silicon comes into contact with even the tiniest amount of the wrong substance, it causes a ruinous chemical reaction. You need crucibles made from the one substance that has both the strength to withstand the heat required to melt polysilicon, and a molecular composition that won't infect it. That substance is pure quartz.[12]

This is where Spruce Pine quartz comes in. It's the world's primary source of the raw material needed to make the fused-quartz crucibles in which computer-chip-grade polysilicon is melted. A fire in 2008 at one of the main quartz facilities in Spruce Pine for a time all but shut off the supply of high-purity quartz to the world market, sending shivers through the industry.[13]

Today one company dominates production of Spruce Pine quartz. Unimin, an outfit founded in 1970, has gradually bought up Spruce Pine area mines and bought out competitors, until today the company's North Carolina quartz operations supply most of the world's high- and ultra-high-purity quartz.[14] (Unimin itself is now a division of a Belgian mining conglomerate, Sibelco.) In recent years, another company, the imaginatively titled Quartz Corp, has managed to grab a small share of the Spruce Pine market. There are a very few other places around the world producing high-purity quartz,[15] and many other places where companies are looking hard for more. But Unimin controls the bulk of the trade.

The quartz for the crucibles, like the silicon they will produce, needs to be almost absolutely pure, purged as thoroughly as possible of other elements. Spruce Pine quartz is highly pure to begin with, and purer still after being put through several rounds of froth flotation. But some of the grains may still have what Glover calls interstitial crystalline contamination—molecules of other minerals attached to the quartz molecules. That's frustratingly common. "I've evaluated thousands of quartz samples from all over the world," said John Schlanz, chief minerals processing engineer at the Minerals Research Laboratory in Asheville, about an hour from Spruce Pine. "Near all of them have contaminate locked in the quartz grains that you can't get out."

Some Spruce Pine quartz is flawed in this way. Those grains, the washouts from the Delta Force of the quartz selection process, are used for high-end beach sand and golf course bunkers—most famously the salt-white traps of Augusta National Golf Club,[16] site of the iconic Masters Tournament. A golf course in the oil-drunk United Arab Emirates imported 4,000 tons of this sand in 2008 to make sure its sand traps were world-class, too.

The very best Spruce Pine quartz, however, has an open crystalline structure, which means that hydrofluoric acid can be injected right into the crystal molecules to dissolve any lingering traces of feldspar or iron, taking the purity up another notch. Technicians take it one step further by reacting the quartz with chlorine or hydrochloric acid at high temperatures,[17] then putting it through one or two more trade-secret steps of physical and chemical processing.

The result is what Unimin markets as Iota quartz, the industry standard of purity. The basic Iota quartz is 99.998 percent pure SiO_2. It is used to make things like halogen lamps and photovoltaic cells, but it's not good enough to make those crucibles in which polysilicon is melted. For that you need Iota 6, or the tip-top of the line, Iota 8, which clocks in at 99.9992 percent purity—meaning for every one billion molecules of SiO_2, there are only eighty molecules of impurities.[18] Iota 8 sells for up to $10,000 a ton. Regular construction sand, at the other end of the sand scale, can be had for a few dollars per ton.

At his house, Glover showed me some Iota under a microscope. Seen through the instrument's lens (itself made from a much less pure quartz sand), the jagged little shards were as clear as glass and bright as diamonds.

Unimin sells this ultra-high-purity quartz sand to companies like General Electric, which melts it, spins it, and fuses it into what looks like a salad bowl made of milky glass: the crucible.[19] "It's safe to say the vast majority of those crucibles are made from Spruce Pine quartz," said Schlanz.

The polysilicon is placed in those quartz crucibles, melted down, and set spinning. Then a silicon seed crystal about the size of a pencil is lowered into it, spinning in the opposite direction. The seed crystal is slowly withdrawn, pulling behind it what is now a single giant silicon crystal.[20] These dark, shiny crystals, weighing about 220 pounds, are called ingots.

The ingots are sliced into thin wafers. Some are sold to solar cell manufacturers. Ingots of the highest purity are polished to mirror smoothness and sold on to a chipmaker like Intel. It's a thriving trade; wafers were a $292 billion industry in 2012.[21]

The chipmaker imprints patterns of transistors on the wafer using a process called photolithography. Copper is implanted to link those billions of transistors to form integrated circuits. Even a minute particle of dust can ruin the chip's intricate circuitry, so all of this happens in what's called a clean room, where purifiers keep the air thousands of times cleaner than a hospital operating room. Technicians dress in an all-covering white uniform affectionately known as a bunny suit.[22] To ensure the wafers don't get contaminated during manufacture, many of the tools used to move and manipulate them are, like the crucibles, made from high-purity quartz.[23]

The wafers are then cut into tiny, unbelievably thin quadrangular chips—computer chips, the brains inside your mobile phone or laptop. The whole process requires hundreds of precise, carefully controlled steps. The chip that results is easily one of the most complicated man-made objects on Earth, yet made with the most common stuff on Earth: humble sand.

The total amount of high-purity quartz produced worldwide each year is estimated at 30,000 tons[24]—less than the amount of construction sand produced in the United States every *hour*. Only Unimin knows exactly how much Spruce Pine quartz is produced, because it doesn't publish any production figures. It is an organization famously big on secrecy. "Spruce Pine used to be mom-and-pop operations," said Schlanz. "When I first worked up there, you could just walk into any of the operations. You could just go across the street and borrow a piece of equipment." Nowadays Unimin won't even allow staff of the Minerals Research Laboratory inside the mines or processing facilities. Any contractors brought in for repair work have to sign confidentiality agreements. Whenever possible, vice-president Richard Zielke recently declared in court papers, the company splits up the work among different contractors so that no individual can learn too much. Unimin buys equipment and parts from multiple vendors for the same reason.[25] Glover has heard of contractors being blindfolded inside the processing plants until they arrive at the specific area where their jobs are and of an employee who was fired on the spot for bringing someone in without authorization. He says the company doesn't even allow its employees to socialize with those of their competitors.

It was hard to check out Glover's stories, because Unimin

wouldn't talk to me. Unlike most big corporations, its website lists no contact for a press spokesperson or public relations representative. Several emails to their general inquiries address went unanswered. When I called the company's headquarters in Connecticut, the woman who answered the phone seemed mystified by the concept of a journalist wanting to ask questions. She put me on hold for a few minutes, then came back to tell me the company has no PR department, but that if I faxed (faxed!) her my questions, someone might get back to me. Eventually I got in touch with a Unimin executive who asked me to send her my questions by email. I did so. The response: "Unfortunately, we are not in a position to provide answers at this point in time."

So I tried the direct approach. Like all the quartz mining and processing facilities in the area, Unimin's Schoolhouse Quartz Plant, set in a valley amid low, thickly treed hills, is surrounded by a barbed-wire-topped fence. Security isn't exactly at the level of Fort Knox, but the message is clear.

One Saturday morning I went to take a look at the plant with David Biddix. We parked across the street from the gate. A sign warned that the area was under video surveillance, and that neither guns nor tobacco were allowed inside. As soon as I hopped out to snap a few photos, a matronly woman in a security guard uniform popped out of the gatehouse. "Watcha doin'?" she asked conversationally. I gave her my friendliest smile and told her I was a journalist writing a book about sand, including about the importance of the quartz sand in this very facility. She took that all in skeptically, and asked me to call Unimin's local office the following Monday to get permission.

"Sure, I'll do that," I said. "I just want to take a look, as long as I'm here."

"Well, please don't take pictures," she said. There wasn't much to see—some piles of white sand, a bunch of metal tanks, a redbrick building near the gate—so I agreed. She lumbered back inside. I put away my camera and pulled out my notebook. That brought her right back out.

"You don't look like a terrorist"—she laughed apologetically— "but these days you never know. I'm asking you to leave before I get grumpy."

"I understand," I said. "I just want to take a few notes. And anyway, this is a public road. I have the right to be here."

That really displeased her. "I'm doing my job," she snapped.

"I'm doing mine."

"All right, I'm taking notes, too," she declared. "And if anything . . ." Leaving the consequences unspecified, she strode over to my rental car and officiously wrote down its license plate number, then asked for the name of "my companion" in the passenger seat. I didn't want to get Biddix in any trouble, so I politely declined, hopped in, and drove off. It was a frustrating encounter for all concerned, but at least there weren't any shovel-toting goons this time.

If you really want a sense of how zealously Unimin guards its trade secrets, ask Dr. Tom Gallo. He used to work for the company, and then for years had his life ruined by it.

Gallo is a small, lean man in his fifties, originally from New Jersey. He relocated to North Carolina when he was hired by Unimin in 1997. His first day on the job, he was handed a confidentiality agreement; he was surprised at how restrictive it was and didn't think it was fair. But there he was, way out in Spruce Pine, with all his possessions in a moving truck, his life in New Jersey already left behind. So he signed it.

Gallo worked for Unimin in Spruce Pine for twelve years. When he left, he signed a noncompete agreement that forbade him from working for any of the company's competitors in the high-purity quartz business for five years. He and his wife moved to Asheville and started up an artisanal pizza business, which they dubbed Gallolea—his last name plus that of a friend who had encouraged him. It was a rough go. The pizza business was never a big money-maker, and it was soon hit with a lawsuit over its name from the E. & J. Gallo Winery. Gallo spent thousands of dollars fighting the suit—it's his name, after all—but eventually decided the prudent course would be to give up and change the company's name. The five-year noncompete term had run out by then, so when a small start-up quartz company, I-Minerals, called to offer Gallo a consulting gig, he gladly accepted. I-Minerals put out a press release bragging about the hire and touting Gallo's expertise.

That turned to be a big mistake. Unimin promptly filed a lawsuit against Gallo and I-Minerals, accusing them of trying to steal Unimin's secrets.

"There was no call, no cease-and-desist order, no investigation," said Gallo. "They filed a 150-page brief against me on the basis of a press release."

Over the next several years, Gallo spent tens of thousands of dollars fighting the suit. "That's how billion-dollar corporations terrify people," he said. "I had to take money out of my 401(k) to defend myself against this totally baseless lawsuit. We were afraid we would lose our house. It was terrifying. You can't imagine how many sleepless nights my wife and I have had." His pizza business collapsed. "When Unimin filed suit, we had just gotten over the Gallo thing. It was the sledgehammer that broke the camel's back. We'd worked on it for five years. It was more than we could handle emotionally, psychologically, and financially."

Unimin eventually lost the case, appealed it to federal court, and finally dropped it. I-Minerals and Gallo separately countersued

Unimin, calling its suit an abuse of the judicial process aimed at harassing a potential competitor. Unimin eventually agreed to pay an undisclosed sum to have the suits withdrawn. Under the terms of the settlement, Gallo can't disclose the details, but said bitterly, "When you get sued by a big corporation, you lose no matter what."

For all the wealth that comes out of the ground in the Spruce Pine area, not much of it stays there. Today the mines are all owned by foreign corporations. They're highly automated, so they don't need many workers. "Now there's maybe twenty-five or thirty people on a shift, instead of three hundred," said Biddix. The area's other jobs are vanishing. "We had seven furniture factories here when I was a kid," said Biddix. "We had knitting mills making blue jeans and nylons. They're all gone."

Median household income in Mitchell County, where Spruce Pine sits, is just over $37,000, far below the national average of $51,579. Twenty percent of the county's 15,000 people, almost all of whom are white,[26] live below the poverty line. Fewer than one in seven adults has a college degree.

People find ways to get by. Glover has a side business growing Christmas trees on his property. Biddix makes his living running the website of a nearby community college.

One of the few new sources of jobs are several huge data processing centers that have opened up in the area. Attracted by the cheap land, Google, Apple, Microsoft, and other tech companies have all opened up server farms within an hour's drive of Spruce Pine.[27]

In a sense, Spruce Pine's quartz has come full circle. "When you talk to Siri, you're talking to a building here at the Apple center," said Biddix.

I pulled out my iPhone and asked Siri if she knew where her silicon brains came from.

"Who, me?" she replied the first time.

I tried again.

"I've never really thought about it," she said.

I can't blame her. Most of us never think about how our high-tech industries depend on sand. Even fewer realize that increasingly, America's twenty-first-century fossil fuel industry does, too.

CHAPTER 6

Fracking Facilitator

On a platform several stories above the North Dakota prairie, a roaring, mud-stained, 1,500-horsepower motor spun a steel rod as thick as a softball bat in an endless pirouette. The rod continued down through about thirty feet of metal housing and then burrowed into the ground. Below the earth, the drill extended for a distance almost twice as long as the Golden Gate Bridge.

Inside an adjoining control room, a fleshy operator whose hard hat identified him as Chuck reclined in a chair surrounded by seven swing-mounted monitors, looking like the king of all video gamers as he tracked the drill's progress. It went about two miles straight down, then turned sideways for another mile. It was chewing its way, at 110 feet per hour, through a second horizontal mile of solid rock. The purpose: preparing all that rock to be hydraulically fractured—a process better known as fracking.

Fracking is about as popular with the general public as kicking puppies, but it's very big business. Nearly 5 million barrels of oil per day, along with immense amounts of natural gas, are being extracted from the fracking fields in North Dakota, Texas, Ohio, and Pennsylvania. Thanks to the fracking boom, which kicked into high gear in 2008, the United States has overtaken Saudi Arabia and Russia to become the world's biggest oil and gas producer.

None of this could happen without sand. America's fracking fields are the latest front to which we have deployed armies of sand to maintain our lifestyle.

Energy companies have known for decades that shale rock formations, such as North Dakota's Bakken Formation, hold huge amounts of hydrocarbons. The problem was extracting them. In conventional oil- or gas-bearing rock, the hydrocarbon molecules flow through pores in the stone into a well, like seawater seeping into a hole dug on a sandy beach. But shale formations are so dense that the oil and gas can't flow through them.

The solution is to fracture—frack—the rock. By shooting a highly pressurized mix of water, chemicals, and sand into a well bore, drillers shatter the surrounding shale, spider-webbing it with tiny cracks through which the hydrocarbons can flow. They need the sand to keep the cracks open, holding fast against the pressure of the surrounding rock that wants to close them back up. In 2000, a Texas oil entrepreneur named George Mitchell refined the technique and married it with the rapidly developing technology of horizontal drilling,[1] with the result that previously unreachable oil and gas became accessible. The rest of the industry soon copied the system, and the fracking boom was on.

American shale gas production totaled 320 billion cubic feet in 2000; in 2016, the number was 15.8 *trillion*.[2] The Energy Information Administration estimates shale gas alone could meet US natural gas needs for the next forty years.

Meanwhile, EOG Resources—a company you may remember by its former name, Enron—started fracking the Bakken in 2006. Since then, North Dakota's annual oil production has nearly quintupled, to over half a million barrels a day. The number of wells has jumped from less than 100 to around 6,000.[3] There's more shale oil coming online in Texas and several other states, and a potential mother lode in California, where environmental concerns have limited drilling—for now.

Every one of those wells needs sand, and lots of it. A single well can use as much as 25,000 tons—enough to fill more than two hundred railroad cars. But like members of a specialized combat unit, frac sand grains need to meet a list of highly specific physical requirements. They must be hard enough to withstand all that pressure, which means they must be at least 95 percent quartz.[4] That eliminates most common construction sand, shrinking the pool to the silica sands used for glassmaking. But frac sand must also have the right shape: small enough to fit snugly into the frack cracks and rounded enough to let the hydrocarbons slide easily around them. Most quartz grains, you'll recall, are angular; there aren't many places where you can find grains with such high purity and low angularity.

The quartz sands under the ground of western and central Wisconsin[5] have just that rare combination. These are ancient grains that were eroded, transported, then buried and uplifted again. Generally speaking, the older a grain is, the more rounded it is, thanks to however many extra million years of having its angles and edges worn down. Wisconsin also happens to have an excellent rail network and relatively lax environmental regulations. And so the fracking boom has sparked a frac-sand boom in the Badger State. Thousands of acres of the state's farmland and forest are being torn up to get at the precious silica below.

In 2010, there were ten frac sand mines and processing plants in Wisconsin; four years later, that number had shot up to 135.[6] The state produced around 25 million tons of frac sand in 2014, worth nearly $2 billion. A sharp drop in oil prices had slowed fracking, and hence the demand for sand, when I visited Wisconsin in 2015, but at the time of this writing in 2017, it was rebounding smartly. Production is likely to continue growing, since oil and gas operators have learned that increasing the amount of sand they shoot into a well increases the yield of oil or gas. New frac sand mines are also being opened in Texas as producers seek sources closer to the oil fields. The Trump administration's enthusiasm for domestically produced fossil fuels can only help the industry's prospects.

Nationwide, the legions of silica sand used for fracking have grown tenfold since 2003.[7] They now dwarf those used for glass-making and all other purposes, including silicon chips. By 2016, total silica sand production stood at nearly 92 million tons per year, almost three-quarters of which was used for fracking. Only 7 percent went to the glass industry.[8]

Many of the locals in the once-quiet, sand-rich Wisconsin counties have profited from the industry's growth. But many others are deeply concerned about the impact of all those mines, and the processing plants, trucks, and other industrial impedimenta that come with them, on the area's air, water, and quality of life. The sand rush has opened deep divisions between its supporters and opponents. "There are family members who aren't talking to each other," said Donna Brogan, a supervisor on the town board of Arcadia, a town of around 3,000 people in western Wisconsin. "There's been huge bad blood."

Chippewa County, in western Wisconsin, is some of the most beautiful farming country you could hope to see. It is miles and miles of gently undulating hills checkered with corn and soybean fields. Lush emerald pastures host lazy little herds of black-and-white cows, dotted here and there with gambrel-roofed red barns and stubby silos. It was late fall when I visited, and the thick swatches of trees along the ridgetops were ablaze with reds and yellows of turning leaves. It was as pastoral-pretty as it gets.

Except, of course, for the sand mines. Across the road from the two-story house where Victoria Trinko lives, a huge tract of that picturesque farmland has been ripped away, leaving a raw brown and yellow weal of exposed earth. Cornfields and bluffs thick with trees lined the edges of what has become a 176-acre industrial zone. Enormous piles of white sand loomed next to a denuded hill-side, the side of which had been sheared away as though with a giant cake knife. A sorting and washing machine, a hulking con-catenation of conveyor belts, ducts, and metal tanks, clattered as it prepared the sand to be loaded into the trucks that rolled in and out, diesel motors grinding.

Trinko's father bought the 80-acre farm where she lives back in 1936. She's been there most of her sixty-nine years, minus some years living in a nearby town. She still mows the grass and cleans the cow barn by herself, driving a little front-end loader. She is proud of having recently shot dead no fewer than seventeen squirrels that were tearing up her bird feeder. The last few years, though, have been the hardest. "If my dad could see the rape of this land, he'd hate it," she said. "It's totally ugly, and it's detrimental to our health."

She loathes everything about the mine—the noise, the truck traffic, the lights at night (the county permits the mine to operate twenty-four hours a day, seven days a week)—but her biggest con-cern is what the dust from the mine was doing to her. In the months after the Chippewa Sand Company opened it in 2011, every time she went outside, she tasted grit in her teeth and felt dust clinging to her face. Her voice got raspy and her throat was always getting sore. Her doctor sent her to a pulmonary specialist, who diagnosed her with asthma caused by her environment, she said. "I've been here all my life, and within ten months of the mine, I've got asthma?"

She now wears a dust mask outside, and has three air purifiers in the house. "I haven't opened the windows in years," she said. "They always say, 'I can do what I want on my land.' But the noise, the fumes, the sand—once it leaves your land, it's on my land."

The Chippewa Sand Company wouldn't let me see its operation. In fact, when I parked outside their gates to snap a few pictures, a worker came out to tell me I wasn't even allowed to stop there. (Remember Unimin, the giant mining company that gave me a similar reception in North Carolina? They're also big players in Wisconsin. In fact, Unimin is the biggest frac sand producer in the world.)[9]

I did, however, get a thorough tour of another sand mine and processing facility operated by Mississippi Sand in neighboring Trempealeau County. Chad Losinski, the plant manager, evidently felt there was nothing to hide.

Losinski is a sturdily built twentysomething with a trace of a Polish accent handed down from his grandparents. I met him in October of 2015, at a moment when oil prices had sagged to historic lows. That had caused a big slowdown in fracking operations around the country, which in turn had idled the Mississippi Sand plant. The place had been fully operational for only two years, in fact, before the slumping demand for frac sand forced them to lay off everyone—some forty-odd people—except for Losinski and one other employee. Such is the boom-bust nature of the energy business. Luckily for me, it meant Losinski had some free time.

Losinski and I clambered up a long flight of steel rungs to reach the top of one of the facility's handful of hundred-foot storage silos. Once we'd caught our breath, we had an excellent view of the whole 231-acre operation.

Losinski pointed to a tree-covered hill that had had a portion sheared neatly away, leaving a forty-foot face of exposed rock, marbled with different colored strata. The company was gradually tearing the whole hill to pieces.

The first step, he explained, is for excavating machines to scrape off the "overburden"—the plants, trees, topsoil, and unwanted miscellaneous rock lying on top of the sandstone that is their target. One reason Wisconsin silica sand is so desirable is because it lies very close to the surface, requiring relatively little digging to get at it.[10] The topsoil is piled somewhere out of the way; it will be needed to help reclaim the land once the mine is tapped out, as required by law. Mississippi Sand has built a huge berm out of the topsoil, which helps block the neighbors' view of the mine. Once the sand is all gone, the plan is to restore the hills; they'll just be about a third smaller than before.

Once the sandstone is exposed, blasting experts drill a grid of holes into it, pack them with explosives, and simply blow a chunk of the hillside to smithereens. The sandstone shatters and collapses in a heap of . . . well, sand and stones. Front-end loaders dump the raw sand into trucks. After the "raw pile" is cleared away, excavators tear off another swatch of overburden and the process starts again, the hill disappearing slice by slice.

Down on the mine floor, the trucks haul the sand a few hundred yards to another pile, from where it's fed into a complicated behemoth of a machine, a forty-foot-high Frankenstein of pipes, tanks, ladders, catwalks, and conveyor belts. A series of belts haul the sand up some thirty feet to a sorting screen, where jets spray it with water to turn it into a slurry.

This sand-water mixture is then pumped onto a series of vibrating metal screens, which separate out first the miscellaneous rocks, then the oversize grains, shuffling these unwanted bits into a waste pile. Once everything bigger than .8 millimeters has been screened out, the remaining slurry is pumped up through corrugated pipe into a kind of upside-down pyramid called a hydrosizer. One hundred jets blast down into the cone, creating a carefully calibrated rising current that carries the lighter grains up and over the top into a trough, while the heavier ones sink to the bottom. By controlling the strength of the jets, you control the size of the grains that sink. The sand that ends up on the bottom is the stuff you keep.

That sand is then run through a series of four attrition tanks—basically giant washing machines that spin the slurry, making the grains grind against one another, washing off silt or other impurities that might coat them. Last stop is a dewatering screen, a mesh of tiny slots measuring .01 millimeters, big enough for water to get through but not sand.

Now partly dried, the sand is fed onto a series of three escalating conveyor belts and sprayed up onto an enormous dune of light-beige sand. Losinski reckoned there was about 120,000 tons in the pile on the day I visited.

(Later, I got a look at processed frac sand grains under a microscope. They were glass-clear, irregularly shaped, but all falling in a narrow range of shape and size, like a bunch of crystalline supermarket potatoes.)

The sand is taken next to the drying plant, a vast warehouse-style building a few hundred yards away. Trucks load the washed sand into a metal hopper that feeds it onto another series of rising conveyor belts that carry it up to a doorway in the dryer plant, some twenty feet above the ground. Inside is a cavernous space, untouched by natural light, filled with another set of machines. The sand gets one more sifting, to filter out any stray rocks that might have gotten in on the journey from the pile, and then is fed through a long cylindrical tank. A series of ducts underneath the tank blows hot air upward, drying the sand, while smokestack-like chimneys whisk away stray silica dust. "That's the bad shit," says Losinski. "That's the stuff you don't want to breathe."

Crystalline silica dust is sharp and jagged, especially when it's freshly formed—like that found at sand mines and processing sites—and it can wreak havoc on the lungs. It's been known for decades that too much exposure can cause silicosis, an especially severe lung disease. In fact, before we set off on our tour, Losinski was legally required to read me a set of warnings including one stating that "prolonged exposure to silica dust can lead to silicosis." When the dryers are running, wearing a respirator is mandatory.

The dangerous dust gets sucked away into a bag and mixed with water to form a paste, which is later buried underground. But despite the safety machinery's best efforts, there are little heaps and hummocks of sand scattered around the plant floor that have sifted out through cracks or bad joins.

Losinski shrugged. "Nothing's perfect."

A final relay of vibrating screens separates the sand into three size grades. Those are then hauled up a hundred feet in bucket elevators, vertical conveyor belts fitted with dozens of fiberglass buckets, and dumped into one of the 3,000-ton silos atop which Losinski and I stood. Trucks drive right up to the silos, fill up, and haul the product to the nearest rail station in Winona, Minnesota. From there, it's off to the fracking fields. The sand that used to make up a Wisconsin hillside will be shot deep into the earth hundreds of miles away in Texas or North Dakota.

There have been small-scale silica sand mines in western Wisconsin for decades, supplying glass factories and foundries. Nobody much minded them. Their impact on the area was manageable. But when the number of mines suddenly mushroomed from a handful to over a hundred in just a few years, locals were taken aback.

"Everyone was really blindsided" by the inrush of frac sand

outfits, said Pat Popple, a retired school principal who has been at the forefront of anti-frac sand mining activism in Chippewa County since the first mine was proposed back in 2008. There are no professional Greenpeace types or idealistic students out here; the industry is opposed mainly by an ad hoc collection of local farmers and homeowners who have educated themselves and each other on the issues.

"I figured they'd be like coal companies and try to pull the wool over people's eyes," Popple said. "We began to realize we were guinea pigs. There had been no studies on the dangers of silica in the air, or what flocculants [chemicals used at the mines] can do to the water. No studies done, and no one asking questions. The county and town board members really didn't know anything about these questions."

"We knew nothing when they first showed up," agreed Dan Masterpole, director of Chippewa County's Land Conservation and Forest Management Department. "We've learned a lot along the way!"

One lesson they've learned is that no one knows for sure what impact the sand mines are having on the region's environment and its residents' health. The mines are just too new. But there are a number of potentially serious risks to be concerned about.

The first is water. The mines need lots of it to create their slurry and to wash the sand; a single mine can run through as much as 2 million gallons per day. The miners get a lot of it from high-capacity wells, which pump more than 70 gallons a minute from underground aquifers.[11] "There's a lot of concern about whether that will affect groundwater and trout streams fed by these headwaters," said Ken Schmitt, a Chippewa County dairy farmer and father of four. He carries around a stack of photos showing damage done by the mines, including several of creeks clouded with beige mud. In 2013, the Mississippi Sand mine was fined $60,000 for failing to prevent rainstorms from washing sand and soil into a nearby creek.[12]

Schmitt is a sturdily built man with black hair fading into white under his red baseball cap, wearing a frayed denim shirt tucked into beltless Wranglers. He grew up on a family farm and has spent pretty much his whole life in the area. He usually votes Republican. When the mining companies started trying to move in back in 2008, Schmitt went to some of the community meetings about them. What he heard alarmed him.

"Every time a mining company spoke, their story always changed," he said. "They'd say, 'You don't have to worry about water or air particulates. You won't even know we're here.' Basically they were lying to us, just saying whatever they figured we wanted to hear so they could get their project in. That kind of pissed me off. I thought, if they're gonna pull this shit, the gloves are off. We're gonna try and stop them." He's become a vocal opponent of the industry at community meetings and to the media.

So far there's no evidence that the mines are seriously depleting groundwater, said Masterpole. Then again, as Schmitt pointed out, "A lot of these problems may not show up until after the companies have left."

There's also the question of what to do with wastewater that has been used to wash and process the sand. Typically the wastewater gets pumped into settling ponds; this is where the flocculants Pat Popple worries about are added in. Flocculants help remove particles suspended in the water, which is good. But they also contain acrylamide, a neurotoxin and carcinogen, which is bad. That compound could potentially leach from the ponds into groundwater or surface water, warns a 2014 report[13] by the Civil Society Institute and Midwest Environmental Advocates, a group based in Madison, Wisconsin. State regulators launched an investigation into the issue in 2016.

Kimberlee Wright, Midwest Environmental Advocates' executive director, also worries about the economic impact of losing all that farmland. "La Crosse has become a global center for biking. There are lots of bed-and-breakfasts and bike trekking companies there now," she said. "Sometimes when the mines are booming, trucks are going by every thirty seconds." Who's going to want to bicycle-tour with that going on?

"We're ninety miles from Minneapolis," says Willem Gebben, a Chippewa County potter whose home is less than a mile from a proposed 1,200-acre sand mine. "Lots of people come here for biking and fishing. No one says, 'Let's hop in the car and go look at a strip mine!' It's a threat to the whole tourism industry."

The most dire concern, though, is over what the mines are putting into the air. The processing plants, heavy equipment, and trucks kick up a lot of dust, including microscopic bits of particulate matter smaller than 2.5 micrometers, known as PM 2.5. When inhaled, particles that size get deep into the lungs, where they can cause or worsen asthma, lung disease, and a range of other ailments. According to *JAMA*, the journal of the American Medical Association, PM pollution is estimated to cause 22,000 to 52,000 deaths per year in the United States alone.[14]

Particles of silica dust, those tiny bits of frac sand that go airborne, are especially worrisome forms of particulate matter. Silica-related lung disease kills hundreds of American workers each year. So it's a real concern for frac sand miners and others who work at the plants or live nearby. A 2012 study of fracking sites by the National Institute for Occupational Safety and Health found potentially dangerous levels of airborne silica in almost half of all the samples it took at eleven different sites in five states—in some cases ten times the levels deemed safe.[15] At the time of this writing, OSHA was drawing up new regulations to boost safety at silica sand mines.

Silica dust is also a worry for Victoria Trinko and anyone else—especially children and elderly people—living downwind from sand mines. According to maps developed by the Environmental Working Group, a nonprofit research outfit, more than 25,000 people in Wisconsin live within less than half a mile of existing or proposed sand mines and related sites, and a similar number in neighboring Minnesota and Iowa. Twenty schools and two hospitals[16] are also in that radius. "There are loads of studies on silica in the workplace, but not in people's homes," said Kimberlee Wright.

The existing evidence is mixed. In 2013, Wisconsin researchers collected sixteen air samples from the fence line around a major sand mine and processing plant in Chippewa Falls, and found that the silica content was far higher than the number set as the chronic exposure limit in California, Minnesota, and Texas.[17] (Wisconsin has yet to set its own air quality standard for silica.) A more recent study published in the journal *Atmosphere*, however, found respirable crystalline silica concentrations near three Wisconsin frac sand mines and a processing plant to be below levels considered harmful.[18] We may not find out who's right for a long time. Symptoms of silicosis can take ten to fifteen years to develop.

Of course, there are local, state, and federal government regulating bodies tasked with making sure frac sand mines operate safely. But because the industry has grown so fast, "the system to permit and regulate them is at best a patchwork of various agencies and can differ substantially from state to state and from locality to locality," according to the MEA report.[19]

In several cases, mining companies that found the rules in a given county too onerous have convinced towns with more lenient environmental attitudes to simply annex the lands they want, enabling them to operate with fewer restrictions.[20] The council of the city of Arcadia, in Wisconsin's Trempealeau County, pulled this maneuver in 2012.[21] Meanwhile, the board of the adjacent *town* of Arcadia handed out permits to more than a dozen mines in a few short years. Locals were so outraged that in 2015 they voted the entire town board out, replacing them with a slate of explicitly anti-sand-mining candidates, including Donna Brogan.

The most important regulatory body in Wisconsin charged with monitoring air and water quality is the state Department of Natural Resources. As of 2014, the department had cited twenty sand mining companies for various rule violations.[22] The department's many critics, however, insist it's too inclined to give businesses the benefit of the doubt. It is certainly under pressure from pro-business interests: In his 2010 campaign, Governor Scott Walker slammed the department for its "out-of-control" enforcement of environmental rules, which he charged were squelching job growth. Walker cut dozens from the department's staff, including at least eighteen senior scientists.[23]

The authorities in neighboring Minnesota, which also has large deposits of frac sand, have taken a much more cautious approach. The state has allowed only a handful of mines to open up so far.[24] In Minnesota's Winona County, just across the Mississippi from Trempealeau County, dozens of protesters were arrested for blocking sand trucks in 2013; the county recently banned frac sand mining and processing altogether.[25]

It's easy to cast the issue in a familiar light: homespun farmers versus land-raping corporations, friends of the earth versus the henchmen of big oil. That's certainly how it looked to a lot of the anti-sand-mining folks in Chippewa. "It's hard not to see a sand mine as an obscenity—a big scar on the landscape," said

Gebben, the potter. "They're tearing up the forests and trees to get at the last bits of oil. It's a crime against future generations."

But the issue looked very different from the kitchen table of the Chippewa County ranch home of Dennis and Darlene Rossa. Five generations of Rossas have lived and farmed on their 700 hilly acres, growing crops in the fields and hunting in the forests. The couple's sliding glass back door looks out on rippling fields of corn rolling away to dense woodlands. Dennis and Darlene's three kids and four grandchildren all live on adjoining farms on the acreage. They all love the land. And in 2013, Dennis and Darlene leased 140 acres of it to a sand mining company.

"We did it for our kids," says Darlene, a redoubtable woman stout of voice, body, and manner, over a piece of homemade pumpkin pie. "It's their future."

"There's no money in farming anymore unless you're really big," Dennis explains, his graying hair neatly combed over the top of his head. Commodity prices are low and competition is fierce; that's why family farms are disappearing all over the country. The Rossas have stayed solvent partly because they're willing to experiment. They've tried raising cattle and hogs, and a few years ago set up a chicken-breeding operation that now produces around a million birds per year.

"At the end of the day, sand is just another commodity, like corn or beans or cattle," Dennis says. In fact, he's expecting the mine to leave the land in better shape than before. "Some of the land they've got is just a knob with trees on top. They'll clear it out, and then we'll have lower, more level land to farm."

Dennis and Darlene aren't greed-blinded corporate patsies. They've just looked at the evidence and their own situation and reached a different conclusion than Trinko or Schmitt. "So many studies have been done," says Darlene. "They haven't got one documented thing to show one person got silicosis from working in these mines." (This is true, although as scientists are fond of saying, absence of evidence is not evidence of absence.)

"We looked into all the health issues," Darlene continues. "Precautions are always taken. As long as you do that, it's fine." She hauls out a white three-ring binder stuffed with maps, documents, and declarations—all the paperwork they had to file to get a permit for the mine. "There's a dust control plan, a high-capacity well plan," she points out, leafing through the heap. "These companies are concerned about water and dust, just like us."

"If there was really something to be concerned about, we wouldn't be doing it here, with our grandkids living here," Dennis chimes in.

Chad Losinski at Mississippi Sand feels much the same. He's lived in Arcadia his whole life, except for the four years he spent at college in La Crosse. Back in 2012, one of Losinski's friends leased his land to another mining company and told Chad he should try to get a job there; the pay was better than the house building he'd been doing.

Losinski didn't know anything about mining, but he was hired anyway. "We start at seventeen dollars an hour, with no skills required, and it goes up from there," he says. That's considerably better than the pay at the local furniture plant, the other big employer in Trempealeau County. As for farming: "Unless you're a big commercial farmer, you can't make it on a small dairy farm. Commodities are worth nothing, and the price of land and everything else is going up. Especially around here—it's a great hunting area. A lot of the land, when it comes for sale, some rich doctor from Green Bay or Milwaukee buys it for hunting." Losinski grew up on a dairy farm, spending his summers baling hay. A couple of years ago, his father sold their herd and came to work at the mine. "He was the hardest-working guy we had," Losinski says. "The money he saved on health insurance alone made it worth it for him to come here."

As for environmental issues: "If there were anything truly concerning in my eyes, I wouldn't be in the industry," Losinski maintains. "I'd probably be sitting on the other side. But I know that we're regulated strictly by DNR for water quality and air, and OSHA for the mining. They're here twice a year for employee safety, to make sure everything is done right. I think it's perfectly safe." What about his neighbors who insist it isn't? "There's just no compromising with them. They just don't want the industry, and that's that."

Indeed, it's just as easy to caricature the anti-sand-mining forces (as some on the pro-sand-mining side do) as a contemptible alliance of paranoid, elitist NIMBY types and local farmers jealous that their own land has no frac sand.

A lot of the complaints about sand mines are that they are a nuisance: they're ugly, they're loud, they spoil the view, they disrupt the peaceful, bucolic feeling of the area. (One woman who lives in a lovely house on a forested hilltop was most upset because a sand mine several miles away spoiled the otherwise perfect rural view. "We haven't entertained on our deck all summer!" she moaned.) All of that is true. But it's also true that those quality-of-life damages come with just about any new kind of economic activity. Every factory, every paved road, every city ever built was birthed amid dust and noise and disruption of whatever patterns of life were there before it, and forever changed the landscape it sat in. For that matter, the lovely farms of Chippewa and Trempealeau counties have been there for only a little over a century. The land they sit on used to be forest. The vast tracts of white pine that once covered much of the state were clear-cut for timber[26] and to make room for agriculture.

That's the course of human history. Cities, highways, factories, modern civilization require tearing up land and displacing people and other living things. It's impossible to get the resources we need to live as we do without disturbing at least some people and doing some harm to—or at least changing—the natural environment. Civilization disrupts the natural world. We disrupt the natural world. But we're not going to go back to living in caves. We're not going to stop cutting down trees or damming rivers or, least of all, digging up sand. The challenge is to figure out ways to do those things that are responsible, sustainable, and limited. We have to do as little of them as we can get away with.

In the specific case of frac sand, though, there's a valid argument to be made that we shouldn't be doing it at all, because fracking itself is especially fraught with serious environmental hazards. There are plenty of reports of fracking operations contaminating aquifers and even causing earthquakes, as well as possibly elevating the risk of cancer and silicosis among people living near them.[27] What's more, society doesn't necessarily need the oil and gas it yields. In an ideal world, it could be replaced with solar and wind power.

That's not an option with other resources, however, especially sand. For its most important uses—concrete and glassmaking—there just isn't a viable alternative (as I'll explain later).

In the meantime, fracking isn't going away, and neither is the demand for Wisconsin's sand. No matter how laughable some of the complaints of Chippewa County homeowners may be, nor how sanguine the sand miners themselves are, there are legitimate reasons to be concerned about the potential for the frac sand industry to overuse groundwater, pollute surface water, and cause silicosis.

All of that is an issue not just for Wisconsin, but for many other parts of America as well. Smaller amounts of frac sand are already being mined in Canada, Texas, and several other states, and there are major deposits in many others. Several other countries are looking into fracking their own shale oil and gas deposits.[28] China has enormous reserves and is expected to start tapping them and mining fracking sand in the coming years.

Dan Masterpole, one of the county officials tasked with making sure government regulations are being followed, is almost painfully diplomatic about the controversy. He's big on "on the one hand this, on the other hand that"-type answers to questions about sand mining's risks to streams or aquifers. Finally, I ask him to just bottom-line it: Should people be concerned, or not?

"People should be concerned because we don't have a significant track record on what the issues are," says Masterpole. "We really have very limited experience. And some of these mining companies also have very limited experience. We're at the beginning of a very long journey."

INTERLUDE
An Incomplete List of Surprising
Practices Involving Sand

As facial treatment: Tired of those wrinkles on your forehead and crow's-feet around your eyes? Here's an easy fix: sandblast your face. That's basically what happens with microdermabrasion, a popular treatment in which a spray of extremely fine silica crystals removes the topmost layer of dead skin cells.

As forensic evidence: The shape, size, and color of sand grains are unique to the geographic area of their origin. Finding out which grains come from where has helped criminal investigators for more than a century. A Bavarian chemist solved a murder in 1908 by identifying the origin of the sand on a suspect's shoes. In 2002, police investigators in Virginia extracted a confession from a suspected killer when they showed him how the sand on his truck matched grains found at a murder scene.

As a replacement for water: The Qur'an instructs observant Muslims to pray five times a day, and to wash themselves each time in a ritual called *wudu,* or wet ablution. Finding water was often tricky, though, in the desert lands where Islam was born—but there was never a shortage of sand. So if there's no clean water to be found, Muslims are permitted to purify themselves instead with a ceremonial dusting of earth or sand, a work-around called *tayammum,* the dry ablution.

As gigantic works of art: At the annual International Sand Sculpture Festival in Antalya, Turkey, artists from around the world mold some 10,000 tons of sand into towering re-creations of everything from the Sphinx to Shrek. Only sand and water can be used, but since the grains from the local beach can be hard to work with, the festival also provides smoother sand from rivers and mountain streams. Antalya's festival is just one of several such around the world, including the US Sand Sculpting Challenge in San Diego. Several hotels in Florida also offer custom sand sculptures as wedding decorations, for as much as $3,000 a pop. Because nothing says "everlasting love" like something made of sand.

CHAPTER 7

Miami Beach-less

It may be the bones of buildings and a tool of the oil and gas industry, but mention the word *sand*, and the first thing most of us flash on is the beach. Who doesn't love those idyllic stretches of coast where the land meets the sea? They're where vacation memories are made and photos taken, where kids build sand castles, teenagers check each other out, lovers stroll in the surf, and indolent adults sip margaritas. They're the global symbol of paradise.

Beaches are also a multibillion-dollar industry. On shorelines around the world, in countries rich and poor, supine armies of sand offer themselves up as tourist attractions that generate livings for millions of people.

That includes most of the residents of Fort Lauderdale, Florida. It's been one of America's prime beach vacation destinations for decades, at least since the 1960 film *Where the Boys Are* made it synonymous with spring-break fun in the sun. But for a place that depends on sun-and-sand-seeking tourists, Fort Lauderdale has a big problem: its beaches are disappearing.

The city has been fighting a defensive battle against nature for many years. The sand that lines its shores is constantly being swept out to sea by wind, waves, and tides. In the natural course of things, that sand would be replenished by grains carried by the Atlantic's near-shore southward-moving currents. That's what used to happen. Today, though, humans have cut off that supply of incoming sand. So many marinas, jetties, and breakwaters have been built along the Atlantic coast in the last hundred years that the flow of incoming sand has been blocked. The natural erosion continues, but the natural replenishment does not.

For decades, Broward County, in which Fort Lauderdale sits, solved its vanishing beach problem by replacing the sand swept off its shoreline with replacement troops dredged up from the nearby ocean floor. But by now virtually all of its accessible undersea sand has been used up. For that matter, the same goes for Miami Beach, Palm Beach, and many other beach-dependent Florida towns. Nearly half of the state's beaches are officially designated as "critically eroding."[1] Nicole Sharp, Broward County's natural resources administrator, summed it up: "We are running out of sand in Florida."

Florida isn't an anomaly. Beaches are disappearing all over America and around the world, from South Africa to Japan and Western Europe. A 2017 study by the US Geological Survey warned that unless something is done, as much as two-thirds of Southern California's beaches may be completely eroded by 2100.[2]

To understand why, you first need to understand how sand gets to the beach in the first place. It usually comes from a combination of sources that vary depending on the local geography. In places with steep mountains close to the shore, like much of the west coast of North and South America, and in deltas like the Mekong in Vietnam, rivers carry sand straight to the shore. On flat coastal plains, like those in the eastern United States, Brazil, and China, some of the sand is left over from ancient river estuaries.[3]

If there are bluffs or cliffs near the water, waves erode them, gnawing off grains that feed the beach. Many beaches also contain biogenic sands—shards of crushed-up shells, corals, and skeletons of marine creatures.[4] That's what makes some beaches pink or extra-white. (Among its many oddly colored beaches, Hawaii boasts a particularly rare one on the island of Kauai called Glass Beach. Much of its sand is made up of millions of colorful pieces of long-eroded glass.) Waves push sand from the ocean bed ashore in some places. And all beaches are fed at least in part by currents traveling along the coast, bringing sand from other areas.

Human beings are interfering with practically all of those processes. Massive coastal development—marinas, jetties, ports—blocks the flow of ocean-borne sand. In many countries, including the United States, river dams also cut off the flow of sand that used to feed beaches. Southern California's beaches have lost as much as four-fifths of the sediment that rivers used to bring them, thanks to dams.[5]

(Human intervention is also changing the flow of sand in ways that reduce territory farther inland. Louisiana loses an estimated sixteen square miles of wetlands every year—a crucial natural defense against hurricanes—because levees and canals on the Mississippi have reduced the flow of sediment that used to replenish them.[6] Egypt's Aswan Dam has done a similar number on the shore of the Nile Delta. China's colossal Three Gorges Dam project is expected to have an even greater impact.)

Sand mining makes the problem worse. Dams combined with upriver sand mining are decimating the supply of replenishing sediment to Vietnam's Mekong Delta, home to 20 million people and source of half that country's food supply.[7] In South Africa, researchers believe sand mining has slashed by two-thirds the flow of river sand that feeds the beaches of the city of Durban. Dredging of near-shore sand to build a railway in Kenya may be eroding some of that country's finest beaches. And in the San Francisco Bay, massive sand dredging may be starving nearby beaches; environmentalists have been battling to stop it for years.

Then there are the places where the beach *itself* is being hauled off. Illegal beach sand mining has been reported all over the world. In Morocco and Algeria, illegal miners have stripped entire beaches for construction sand, leaving behind rocky moonscapes. Thieves in Hungary made off with hundreds of tons of sand from an artificial river beach in 2007. Five miles of beach was stripped down to its clay foundation in Russian-occupied Crimea in 2016. Smugglers in Malaysia, Indonesia, and Cambodia pile beach sand onto small barges in the night and sell them in Singapore.[8] Beaches have been torn up in India and elsewhere by miners seeking rare minerals like zircon and monazite that are found in minute quantities amid the quartz grains. Even farmers in Scotland and Northern Ireland have been known to steal beach sand to improve the quality of their soil.

Perhaps the most notorious heist was in Jamaica, where over the course of a few weeks in 2008, thieves made off with a quarter-mile stretch of lovely white-sand beach near the town of Coral Springs. A planned resort under construction on the beach was brought to a standstill; police officials speculated the 500-odd truckloads of grains were sold to rival developers elsewhere on the island, possibly with the collusion of local police. Five men were eventually charged with the crime, but the case was dropped when a key complainant, an executive with the Coral Springs development company, refused to testify, saying he had received death threats.

In some places, beach sand mining is perfectly legal, if ill advised. Beginning in the 1920s, six operations mined sand along the

California coast. Five of them were finally shut down in 1989 over concerns about shoreline erosion. The last one, owned by the Mexican building materials giant Cemex, was still sucking up sand from a beach near Monterey as recently as mid-2017. After years of pressure from environmental groups and state regulators, however, Cemex has agreed to shutter it by 2020.

Government officials in Puerto Rico have had to restrict beach sand mining because so many grains were being taken to build tourist hotels that the very beaches those tourists came for were disappearing.[9] Many other Caribbean islands have historically used beach sand as their primary supply for making concrete. And some of the poorer islands sell their beach sand to their wealthier neighboring islands that need it to fatten up *their* beaches.

Mining sand from beaches and dunes was for decades one of the primary industries for the 1,600 inhabitants of the tiny Caribbean island of Barbuda. In 1997, a judge ordered the mining to stop because of the widespread environmental damage, but the ban didn't last long.[10] "Would you prefer to appear to be protecting the environment and then have your people go hungry?" the chair of the island's council said to a local reporter in 2013.[11] It's a question that applies in many parts of the world. As of this writing, the industry's future on Barbuda was uncertain; a massive hurricane in September 2017 forced the entire population to evacuate. The storm's damage would likely have been less intense had the islanders not demolished so many protective dunes.[12]

Meanwhile, thanks to climate change, the seas are slowly rising, encroaching onto shorelines. Add rising seas to shrinking beaches and you have a serious problem worldwide. As the ocean draws ever closer to buildings and roads, it poses a major and growing threat to lives and property. It also means big business for Bernie Eastman.

Eastman is a professional beach builder. On a sunny day in January of 2016 in Fort Lauderdale, he took me for a spin in a sort of all-terrain golf cart to show me the project he was then working on: Broward County's latest, $55 million effort to artificially bulk up its shores. "Beach nourishment" is the officially preferred term.

We rode along an expansively wide stretch of creamy-white beach for a mile or so, the Atlantic on one side, villas and hotels on the other, until the sand abruptly dropped off in a miniature cliff about five feet high. From the base of this declivity, the shoreline shrank into a narrow belt of tawny sand.

The tawny grains, replete with seaweed, shells, and bits of coral, were the ones that nature put there. The white ones, unadulterated with even a speck of foreign matter, were deployed by Eastman. Those grains had, just a few days ago, been dug out of a hole in the ground over a hundred miles away in Florida's interior. Eastman was dumping thousands of tons a day of them onto the shore to fatten up the beach. "When we started, waves were lapping up against people's houses," said Eastman.

Having blocked the natural processes that used to feed beaches, people are now replacing them with artificial ones. Beach nourishment, also known as beach replenishment, has become a major industry. More than $7 billion has been spent in the United States in recent decades on artificially rebuilding hundreds of miles of beach nationwide. Almost all of the costs are covered by taxpayers; much of it is overseen by the federal US Army Corps of Engineers. Florida accounted for about a quarter of the total, according to researchers at Western Carolina University. Hundreds of beaches in other countries around the world are also regularly restored with sand brought in from somewhere else.

It's a lucrative business. Eastman is a compact, middle-aged guy with a weather-beaten face adorned with a scrap of white beard and mustache. He tops it all off with a cowboy-hat-shaped hard hat. Eastman's father was in the construction business, and Eastman and his three brothers grew up greasing the trucks. By his own account, Eastman barely graduated from high school. But he took a bunch of night courses to learn things like project estimating, and started his own contracting business in 1994.

His company did all kinds of contracting work, including a little beach renourishment, until the real estate market crash in 2006. Eastman realized that he would do better to rely on the steady forces of erosion and the government funding earmarked to fight it than to tie his fortunes to the vicissitudes of the real estate market. "When the market dried up, we reinvented ourselves," he says. Today Eastman Aggregate Enterprises does nothing but beach nourishment, all over Florida and in neighboring states. Eastman has five of his own trucks and forty-plus people working for him. His company hauls in about $15 million per year.

All told, Eastman Aggregate would dump a million tons of new sand on Broward's beaches over the course of several months. The grains are mined from an inland quarry a couple of hours drive away. Trucks haul that sand down the highway, squeeze their way in between the villas and hotels, and dump it on the shore. Excavators load the freshly delivered sand into hulking yellow dump trucks, which ferry it to the edge of the renourishment zone. Small bulldozers then push the grains into place, extending an evenly proportioned beach out into the surf. "We're putting ten thousand tons a day into the ocean," said Eastman, with no small pride.

Hauling and placing sand with trucks is both considerably slower and far more expensive than the more common method, which is to dredge sand from the sea bottom and blast it onto the shore through floating pipes. The problem is that over the last four decades since beach nourishment began in earnest, Broward

County has used up all the sea sand it is legally and technically able to lay its hands on. Nearly 12 million cubic yards[13] of underwater grains have been stripped off the ocean bottom and thrown onto Broward's shores. There are still some pockets of sand on the seabed, but dredging them is forbidden because it could damage the coral reefs they sit next to. The same goes for Miami-Dade County to the south. In Palm Beach County to the north, most of what little sea sand remains was being sprayed onto its slenderized beaches during my visit in 2015.

There is lots of sand left off the coasts of three other Florida counties farther north. They haven't worked their beaches quite as hard as the tourist meccas to the south, and the continental shelf up there extends further out before dropping into the deep ocean, giving them a larger area to dredge from. Miami-Dade has asked for help, but the northern counties have so far refused to share. They don't want to find themselves in Miami's position thirty years from now. "I'll fight the Army Corps taking even one grain of sand from our beaches," thundered a state senator from the region in 2015.[14]

Desperate Miami-Dade officials are now talking about importing foreign mercenaries, in the form of sand from the Bahamas. This island nation, which is less than two hundred miles from Florida, does have beautiful sand and recently agreed to allow it to be exported. The sticking point is an American law, passed at the urging of the dredging industry, that bans federal funding for beach nourishment projects that use nondomestic sand. And since the federal government typically covers more than half the cost of such projects, Bahamas sand is pretty much off the table. A few years ago, Broward County was even considering using artificial sand made from recycled glass; that turned out to be technically plausible but ridiculously expensive.

Which leaves many towns in southern Florida no choice but to dig their sand from inland quarries and haul it to the coast one roaring, diesel-spewing truck at a time. Tourists and locals hate the noise and traffic, and county officials hate the extra cost, which can be easily double that of dredged sand. But it does have some advantages. The inland mines, with their elaborate sorting and washing machines, can deliver sand of a precise spec—the exact size, shape, and color county officials deem appropriate for the beach.

Beach town residents and tourists alike are very particular about the color and consistency of their beaches. The sugary white-sand beach has become the global standard of perfection, and any resort falling short of it loses points. (That's nothing compared to the fussiness of Olympic beach volleyball players. To make sure their bare feet come into contact only with grains of just the right size and shape, sand was brought in from Hainan Island for the 2008 Beijing Games, and from a quarry in Belgium for the 2004 Athens Games.)[15]

"You pump sand from the ocean floor, you don't know what you're getting," said Eastman. That's not exactly true; sea sand is examined closely to make sure it is suitable for a given beach before the regulatory agencies will allow it to be dredged for nourishment. But land-mined sand can be sorted, sifted, and cleaned to a uniform standard. The grains that Eastman was emplacing were all about the size of a salt grain, all the same silver gray, unadulterated with stones or shell fragments. Their color was approved using the Munsell color order system, a visual index of hues created in 1915. The sand is tested at the mine, at every 3,000 tons, and every 500 yards on the beach after it's in place to make sure it's up to spec. The waves will gradually mix in shells and other organic matter, so in a few months it won't look as obviously artificial as it does now.

Whatever you may think of the process, the beach Eastman is building is magnificent: miles and miles of soft, thick, even sand. On a stretch completed just a few days earlier, retirees were lounging in sun-facing deck chairs, kids were building elaborate castles, and couples were strolling barefoot. You'd never know the sand came from a giant pit many miles away, and that this beach was open water just a couple of weeks ago.

At the same time, renourishment is the embodiment of a Sisyphean task. This particular beach is only expected to last about six years before it needs more upkeep.

Most people think of beaches as timeless parts of the natural world, as places to connect with the elements of sea, sky, and earth. In fact, though, many beaches—including some of the world's most famous—are artificial constructs, engineered environments built for profit. In such places, the original shoreline, the wild coast in its prehuman state, has been obliterated, buried under imported sand. In Broward County, they make no bones about it. "Beaches are a form of infrastructure," said Sharp. "You pave your potholes, we pave our beaches with sand."

For most of human history, beaches weren't places to relax, but to work. The sandy shores were where fishermen launched their boats and cleaned their catch, where small traders unloaded their cargo. Coastal people built their homes a safe distance from the unpredictable weather and waves of the shoreline, often facing away from the sea for added protection.[16] "When Europeans and Americans first settled the coasts, they largely ignored, indeed avoided, what are today's most coveted stretches of shore," writes historian John R. Gillis in *The Human Shore*, an account of our changing relationship with our coasts. "The beach was used for landing but not for settlement. Its featureless barrenness was not only inhospitable but repulsive."

That began to change in the early eighteenth century, when it became fashionable for ailing English elites to visit seaside resorts for treatments with the supposedly curative powers of cold seawater. "They came not to swim but to bathe, and they were assisted in that activity by so-called bathing machines, cabins on wheels that transported them across the beach and into the water, where, with the assistance of hired attendants, women and men alike dipped into the sea as part of their mental and physical cures, which also included drinking sea water, considered at the time medicinal," writes Gillis. Few people knew how to swim in those days, and beaches "were more associated with invalids than athletes, with diseased rather than healthy bodies."[17]

Salt water gradually lost its reputation as a cure-all, but beach tourism developed into its own industry. "1820s-era England is responsible for a turning point in the history of seaside resorts, as this was when the first major bathing establishments were constructed for the specific purpose of bathing, relaxation, and play,"[18] writes University of Florida scholar Tatyana Ressetar in her master's thesis about the history of such places. The popularity of beaches grew through the late 1800s among the burgeoning middle class, with their newfound leisure time, and as railroads made the shores accessible to lower-class city dwellers who previously had no way to reach them.[19] "Once the railroad routes to the beach were complete and cheap excursion and one-day tickets were sold, [the urban poor] took more advantage of the new opportunity, and began to change the leisure industry forever," says Ressetar.

Swimming became a more popular pastime. Swimwear generally consisted of underwear or nothing at all—to the extent that in

Australia, officials concerned about public decency banned beach swimming during daylight hours. Such worries were soothed by the introduction of bathing suits suitably modest for both men and women—typically a neck-to-knee cotton or wool outfit. Los Angeles and its neighboring cities had ordnances requiring such thorough coverage for both men and women; topless men were still being arrested as late as 1929.[20]

Still, the beach itself continued to be viewed with suspicion. Seaside towns like the New Jersey Shore's Atlantic City and the French Riviera's Nice built boardwalks and piers so that visitors could enjoy the sights of the shore without having to actually set foot on its smelly, seaweed-strewn sands.

In time, resort owners took to clearing off the unsightly flotsam and jetsam, and banished fishermen to less desirable areas so that visitors could stroll on the sand itself. Hotels and villas proliferated as the growing urban working class took to seaside vacations. The rich began building private seaside mansions, and the middle class copied them on a smaller scale, until by the 1930s there were seaside towns all over Europe and North America. The rise of the automobile and post–World War II prosperity brought unprecedented numbers to the beach, more and more of whom chose to retire there as time went on.

The beach came to symbolize a refuge from the hectic pace of the modern world, a place of pure leisure. A seaside vacation doesn't require visiting ruins or churches, or standing in lines for rides, or really doing anything at all. There are plenty of activities if you're in the mood. You can run or swim or surf or collect shells or dig holes if you want to, but you don't have to do anything except sit around, in comfort and at ease. The open sands are a blank canvas. "The beach was created ex nihilo, providing neither the sense of place nor the sense of history associated with other vacation destinations, like the country village. The beach thus began as a non-place, a void, and it has remained so ever since. From the start its emptiness, its artificial desertification, has been part of its appeal," writes Gillis. "The appeal of the beach lies in the fact that it excludes all that is 'workful.' Its true relation to nature and history must always be concealed, for it functions in modern culture as a primary place of getting away, of oblivion and forgetting."[21]

The growing appeal of coastal land in sunny climes does a lot to explain the rise of the state of Florida. Originally just a swampy, disease-ridden protrusion from the US mainland, it was largely avoided by sensible folks until real estate developers began selling it to denizens of the overcrowded cities of the Northeast as a place to escape the winter cold. In the 1890s, Standard Oil cofounder Henry Flagler decided to create a new playground for the eastern elite on the South Florida shore, in the little town of Palm Beach. He brought in a railroad, tore out the indigenous mangroves to make room for imported coconut palms, and built luxurious hotels (as well as the town of West Palm Beach, to house his workers). Soon he extended the railway to a parcel of scrubland at the state's southern tip where there was a tiny settlement called Miami. It swelled from a town of 1,681 souls in 1900 to an urban sprawl (built with concrete, of course) of over 200,000 by 1930.[22]

Flagler's railroad also spawned other towns along what had been a barely inhabited coastline, and swelled existing ones, including Fort Lauderdale, a tiny burg named after a stockade from which American soldiers had once fought Seminole Indians. Before the trains appeared, notes the county's official history, the area was mostly swampland, accessible "to only a hardy few."[23] As the population grew, the town became part of a newly constituted Broward County, named after former state governor Napoleon Bonaparte Broward, an energetic swamp drainer.

(Broward was also an unabashed racist who called for the eviction of all black people from the state.[24] He must have been spinning in his grave when, decades later, during the 1960s civil rights movement, black residents of the county bearing his name staged a series of wade-ins on all-white beaches. Despite the efforts of Fort Lauderdale officials in court, and crowds of angry whites on the scene, the campaign succeeded in legally desegregating the beaches.)[25]

The real estate boom in such a wide-open field, the prospect of creating custom-built cities and making tons of money while doing it, drew in investors, moneyed visionaries, and hucksters from around the country—including our old friend Carl Fisher, the man who got the Lincoln Highway built.

In 1916, Fisher opened another continent-straddling road, the Dixie Highway, linking the Midwest to Florida. As intended, it brought more visitors to the state. Fort Lauderdale opened its first tourist hotel in 1919.[26] Fisher was after a bigger prize, though. He set his sights farther south, buying up hundreds of acres of sand-fringed swampland near Miami. "Fisher's Folly was a vermin-infested swamp on the ocean side of Biscayne Bay," writes T. D. Allman in *Finding Florida*. "This boggy wilderness, he decided, was going to be to people with automobiles what Palm Beach was to those with private railroad cars."[27] Fisher tore up the mangroves, dredged millions of tons of sand and mud up from the bay, filled in his land until it was solid enough to be built on, and proclaimed it Miami Beach.

It was an audacious bit of mega-scale landscaping, but not an unprecedented one. Some of the world's most famous beaches were similarly created or expanded with a mass relocation of sand from elsewhere. A century ago, Hawaii's Waikiki Beach was a narrow ribbon of sand fringed by marsh; it was beefed up to its current expansive size with grains barged in from other Hawaiian islands,[28] and at one point in the 1930s with sand shipped from California. Today it still requires regular renourishing. Many of Spain's Canary Island beaches were just rocky coastlines until developers dumped tons of sand imported from the Caribbean and Morocco on them.[29] Half a dozen of the beaches in Barcelona, Spain, were manufactured for the 1992 Olympics. It is such an established practice that Paris now builds a beach on the Seine for a few weeks each summer.[30] (Meanwhile on the southwestern French coast, locals have been protesting for years against beach sand mining.)

Fisher bedizened his prefab paradise with a fancy hotel and casino, not to mention a herd of cows to supply fresh milk to the guests and a baby elephant to pose for pictures with them. He opened a yacht harbor and polo fields and hosted speedboat races. Business boomed. By 1925, Fisher's Florida holdings were valued at over $100 million—more than $1.3 billion in today's dollars.

But the following year, a hurricane packing 130-mile-per-hour winds bore down on southern Florida. The howling winds and surging waves smashed the walls of Fisher's hotels and flooded their lower floors, sweeping smaller buildings away completely. Scores of people were killed. All those northerners and others who had been rushing into Florida suddenly had second thoughts, and the real estate market swooned. Three years later, the stock market crashed, and with it Fisher's fortunes. He died ten years later, a near-penniless alcoholic.

Miami Beach, of course, went on to a much more glamorous and lucrative future, and so did Broward County just to its north. Fort Lauderdale was famous for years as America's spring-break capital, a title it has worked hard to shed since 1985, when a record 350,000 students overran the place. The city now prides itself more on its yachting facilities.

Today the beach-based permanent vacation lifestyle Fisher did so much to help popularize is central to Florida's economy and identity. Tourism is the top industry in the unsubtly nicknamed Sunshine State. Broward County alone pulls in 14 million tourists to its beaches every year, reaping some $6 billion. Statewide, 71 million tourists visit the state each year; some 23 million of them come primarily to spend time on its beaches, generating more than $41 billion in direct and indirect revenues, according to a 2000 study.[31]

Flagler and Fisher opened the way to southern Florida. But it was the interstates that really brought the masses. The I-95 interstate highway funneled people straight down from the big cities of the East Coast, the I-75 brought them in from the Midwest, and the I-10 from everything west of the Florida Panhandle.

These developments interlock, like sand grains interlocking to form concrete. The glamorization of the sandy beach gave rise to cities like Miami Beach and Fort Lauderdale. Roads built of sand made it possible for people to drive to them. Concrete made it possible to build whole cities in the middle of nowhere to house them all. Later, concrete built the vast theme parks—Walt Disney World, Universal Studios—which attracted even more people. Sand abetting sand abetting sand.

In any beach resort, sand underpins the whole tourist economy. Sun and sea are great, but without a soft sandy beach, at best you've got one of those moderately charming Mediterranean towns where you can sunbathe on a rock or a concrete breakwater. That's not a draw for millions of tourists. Sand transforms a place that's merely hot and adjacent to the sea into a universally desired destination. Add sand, and suddenly even the muggy, malarial coast of South Florida is worth a fortune.

Countless other places around the world, from the Black Sea to the Bahamas, depend on the money brought in by outsiders seeking that magical combination of sun, sea, and sand. Hawaii would just be a big pineapple plantation without beaches. Fiji's gorgeous shores attract $1 billion worth of tourists each year—more than the tiny Pacific nation makes from its top five exports combined.[32]

Increasingly, though, beaches are also coming to be valued for something else that might prove even more important than tourist revenue. These seaside armies of sand are a powerful protective force for the people living near them. Beaches are bulwarks that can protect lives and property from storms and rising seas in our climactically imperiled world. Coastal protection has become one of the main justifications for beach renourishment, and with good reason.

All the while that climate change has been accelerating, more and more people have been settling on the shore. Especially since the 1960s, Americans have flocked to coastal communities not only for vacations but to live full-time. Ports, fishing towns, and empty spaces along the coasts have turned into seaside suburbs and retirement communities. Between 1990 and 2010, a Reuters analysis found, about 2.2 million new housing units were built near America's shores, many of them in areas considered most imperiled by sea rise. A third of them were in Florida.[33]

If you think that's a bit crazy, consider this: The US government encourages it. Washington subsidizes local governments and homeowners who build in imperiled coastal areas to the tune of billions of dollars[34] in the form of insurance guarantees, disaster bailouts, and other protections.[35] Taxpayer-funded beach nourishment also has the perverse effect of shoring up property values, a recent study found.[36]

There's $4 billion worth of upland infrastructure—hotels, homes, and other structures—just on the barrier island off

Broward County's shoreline. All told, an estimated $1.4 trillion worth of real estate lies along America's shores. All of it—along with countless billions more in coastal communities in other countries—is endangered by the rising seas, more powerful storms, and more frequent "king tides" spawned by the changing climate.

America's densely populated eastern seaboard is already seeing increased flooding,[37] not to mention more severe storms. When superstorm Sandy assaulted the East Coast in 2012, it killed 159 people, damaged or destroyed at least 650,000 homes, and caused some $65 billion in damage.

The storm's impact was at its most severe in areas where beaches had eroded, leaving little or no buffer between cities and the raging wind and waves. On the other hand, renourished beaches in New York and New Jersey prevented an estimated $1.3 billion in damages that would have been caused by Sandy, according to the US Army Corps of Engineers.[38]

Sand dunes, it turns out, are also good defenses. For decades, developers have bulldozed sand dunes to create more usable beach space and unobstructed views for hotel guests and condominium dwellers. But over time, experience has proven that natural dunes, when they are left in place, can be very effective at protecting those buildings. "Post-Sandy, every coastal community has changed its opinion on dunes," says Nicole Sharp. "People really recognize the storm protection they provide." Naturally occurring sand structures defending human-made ones.

Given both their economic and defensive importance, protecting beaches is of the utmost importance to Florida, as well as the many other places around the world that have linked their fates to the shifting sands of their shorelines. In many places, beaches are reinforced by "armoring" them with stone or concrete seawalls or groins, solid structures sticking out from the beach. These have fallen largely out of favor, though, since research has found they often end up worsening erosion over time by strengthening currents, reflecting waves back onto beaches, and blocking the incoming flow of natural sand.

Which brings us back to beach nourishment. Beaches have been artificially bulked up with sand from elsewhere since at least as far back as a Coney Island project in 1922. The practice came into widespread use in the mid-1960s after a particularly potent storm frayed New Jersey's beaches.[39] Broward County has been doing it since 1970.[40] Remember the "inexhaustible" supply of beach sand that drew home builders to New York's Long Island? Those beaches too have had to be renourished. It's now standard practice all over the world. (It's not always easy, though. One proposed replenishment project in Mumbai had to be put on hold in 2016 because city officials couldn't find enough sand.)

Nourishment, though, is not a cure for beach erosion; it's a treatment, one that must be repeated regularly. Few replenished beaches last longer than five years or so before they have to be fattened up again. Dozens of Florida beaches have been nourished again and again by now, some as many as eighteen times. More than a quarter of a billion cubic yards of sand have gone into the effort. New Jersey's Ocean City Beach has been replenished thirty-eight times, and Virginia Beach, Virginia, more than fifty times.[41]

It's an expensive process. Nourishing a beach can cost up to $10 million per mile.[42] Broward County alone spent more than $100 million replenishing its twenty-four miles of beach in a multiyear project launched in 2015. More than a few individual beaches, such as Atlantic City, have already racked up tabs of well over $100 million by themselves.

And the costs will only keep rising. Andy Coburn, a coastal scientist with the Program for the Study of Developed Shorelines at

Western Carolina University, calculates that the cost of sand for nourishment has multiplied eightfold since the 1970s. It's now more than $14 per cubic yard, a figure he projects will continue to rise as demand increases and the most accessible sand gets tapped out.

Sure, it's expensive, but beach nourishment, the argument goes, more than pays for itself considering what tourism brings in to local, state, and regional economies. As a straight financial proposition, this is irrefutable. But there are other costs involved that can't always be priced in dollars.

Artificial beach building can damage the environment profoundly. Academics and environmentalists have documented how this happens. Geologists Harold Wanless of the University of Miami and Orrin Pilkey of Duke University, among others, have been sounding the alarm for many years about the impacts of beach nourishment on marine ecosystems and habitat. But you'd be hard-pressed to find any critic more dedicated than self-appointed activist Dan Clark.

Dan is a chubby, ruddy-faced man with a long red ponytail who founded and heads an organization, Cry of the Water, devoted to protecting the local coral reefs. Clark was raised in Wisconsin horse country, where his great-grandfather once trained zebras and horses for the Ringling Brothers circus. When Clark was about eight, he moved with his mother to Broward County, where he discovered his life's passion, scuba diving.

"The reefs where I learned to dive in the seventies have been buried," he lamented. "The last of the good stuff is right here."

Clark and his wife, Stefi, scratch out a living taking care of vacant vacation properties and other odd jobs. "We'll scrub boats, toilets, whatever it takes to make a buck," Dan said. For the last two decades he has done just about everything short of throwing himself in front of a bulldozer to stop beach nourishment in

Broward County. He has filed lawsuits, lobbied government officials, made a nuisance of himself at community meetings, and made sure the local media hears from him every time the subject comes up. "I've been fighting for nineteen years," he said proudly.

There's no question that there are ways beach nourishment can harm wildlife and the environment. In Florida, the victims everyone seems most concerned about are the endearing sea turtles that clamber out of the Atlantic from March to October to lay their eggs on the beach. Broward County allows beach nourishment to be done only outside of those months so as not to impinge on the turtles' nesting season.

The new sand also has to match the characteristics of the naturally occurring sand, lest it put off the turtles. If the grains are too sharp, the turtles might avoid them; too dark, and the beach will get too hot and damage the eggs. The slope of the beach also can't be too steep, or the turtles might not be able to climb them. Eastman's crew even tills the sand with huge rakes after it's been put in place, to make sure it's not too hard-packed for the turtles to clamber over. But even with all this consideration, a handful of endangered loggerhead turtles were accidentally killed in 2015 by a trawler that was sucking up sand to spray onto Palm Beach County's shore.

Beach sands are home to a multitude of other creatures, above and below sea level. Besides the obvious visible ones—clams, crabs, birds, plants—they also shelter all kinds of nematodes, flatworms, bacteria, and other organisms so small they live on the surface of individual sand grains. Despite their tiny size, many of these creatures play an important role in the ecosystem, breaking down organic matter and providing food for other creatures, including fish.[43] Dumping thousands of tons of imported sand on top of these organisms can be lethal to them. A 2016 University of California study found the population of marine worms and other invertebrates on San Diego beaches fell by half after a beach nourishment project.[44] Another recent study in South Carolina found major drops in populations of bugs, worms, and other organisms living on the ocean floor in areas that had been dredged for beach nourishment.[45]

The coral reefs that lie just off the southern Florida shore are another contentious issue. They have been directly damaged in the past by dredging ships trolling for sand—which is why Miami-Dade and Broward Counties no longer allow that to happen. But the most stubborn problem is turbidity, the clouding of water by stirred-up sand. Sand suspended in the water can block light from reaching the corals, and when the grains settle, they can suffocate the reefs and whatever creatures are living on them. Clark showed me a sheaf of laminated underwater photos he's taken over the years. One batch shows corals covered with a thick layer of silt, as if they'd sat for years in a long-unvisited attic. "Even what doesn't get buried gets affected by the silt and sediment," said Clark. In 2016, another nourishment project under way in neighboring Palm Beach County had to briefly shut down several times because the water's turbidity levels rose too high.

Dredging sand from the ocean floor generates the most turbidity, but even the grains hauled in by truck cause some. No matter how it is delivered to the beach, some of that freshly placed sand—more loosely packed than natural beaches—inevitably gets swept into the water. In Southern California in 2016, sand from a re-nourishment project drifted into the mouth of the Tijuana Estuary, clogging it so badly that when it rained in Tijuana, the estuary filled up with fish-killing sewage.[46]

Contractors like Bernie Eastman are required to hire third-party consultants to regularly check the turbidity levels they stir up. That's not good enough for Clark. He believes consultants cherry-pick their samples, taking water from the edges of the sediment plume, rather than from its center, where the sand is most densely concentrated. "You can make the case that a football field is white if you only take samples from the lines," he likes to say.

"The consultants have lots of pressure on them to keep the project running," added Ed Tichenor, an environmental activist who does more or less in Palm Beach County what Clark does in Broward. "They're getting paid eight hundred dollars a day. If they keep shutting the project down, they won't have a job."

This is a consistent problem facing anyone trying to figure out the impacts of any process that affects the environment in complex ways. There's always the question: How reliable is the data? Who gathered it? What is their motive? If you're suspicious enough, practically no one can be trusted not to skew the results.

Clark does his own testing. More than once, he said, he's put on an orange safety vest, tucked his hair under a hard hat, and bluffed his way through a work crew to take a sample of sand right off a truck to see if it meets the county's specifications. Sometimes he goes out in his fishing boat to take samples of seawater to check its turbidity. Most often, he takes samples from just off the beach. I followed along one day as he and Stefi set out with a bagful of empty plastic water bottles, Sharpies for labeling, and a tiny wrist-mounted GPS unit to record the location where they took each water sample.

We strolled down to a section of newly placed beach that Eastman's crew had completed a couple of days earlier. Clark waded into the water, getting his boots and pants cuffs soaked. He filled up one of the water bottles and brought it back to show me. It was so clouded with silt that it looked like chocolate milk. "They're not washing it. Not nearly enough. Thing is, they have the ability to do it, but it costs," said Clark.

About half a mile farther south, the renourished area ended and we were back on native sand. Dan filled up another bottle. The water in this one was almost completely clear. He shook it to show me how quickly the sand settled back to the bottom, leaving the water clear again. The samples from the nourished areas were still a semi-opaque brown, and there was a film of bubbles forming on top, like the head on a beer. "That might be phosphates causing those bubbles," says Clark. Another potentially damaging contaminant.

The only real way to completely avoid the pitfalls of beach nourishment while also saving coastal cities is to move those cities inland. Retreat is a radical notion, but it's one that a number of researchers are actively promoting.

It's hard to imagine that actually happening, though. So far we have chosen defense over retreat. Miami Beach is investing $400 million in building seawalls, elevating streets, and installing pumps to combat an anticipated increase in flooding caused by the rising ocean. Around the world, coastal cities like Jakarta, Indonesia, and Bangkok, Thailand, are spending billions on giant seawalls and other protective measures.

In retrospect, it was obviously folly to build so much so close to the ocean's edge. But now there are millions of people and billions of dollars worth of buildings in place; how could we undo all that? No one knows, and few are asking. Which leaves us more or less obliged to keep rebuilding beaches, both as defenses against the ocean and magnets for tourists. The question is, how long can we keep it up before either the money or the sand runs out?

Mike Jenkins is a lean, fortyish coastal engineer with Applied Technology & Management, an engineering firm specializing in seaside structures like marinas and artificial islands. He has also overseen lots of beach nourishment projects. He knows far better than most what the challenges are.

"At some point, it is unsustainable," he said in a conference room of his company's headquarters in West Palm Beach. "Now, that might be a hundred years away, or two hundred years away, but at some point you're going to dredge everything that you can get your hands on." We can extend that time by reengineering some of the man-made inlets and jetties that block the sand's flow, he said, but in the long run there's an even bigger problem. "The ultimate source of supply is rivers. When you start talking decades out, the fact that all the rivers are dammed means that that supply of sand isn't there anymore. But it could take a hundred years before you start noticing.

"The demographics are such that people are moving to the coast. Infrastructure is being built on the coast," Jenkins said. "Now, is that smart? Probably not. But we're doing it."

We keep on building our castles of sand, heedless of the incoming tide.

INTERLUDE
7,500,000,000,000,000,000

That's the best estimate available as to how many grains of sand there are on the world's beaches—7 quintillion, 500 quadrillion. Or 7.5 billion billion, if you prefer.

That magnificent statistic comes courtesy of Howard McAllister, a researcher at the University of Hawaii. He came up with it by guesstimating the world's beaches are covered an average of 30 meters deep by 5 meters wide with sand grains averaging 1 cubic millimeter in volume. He might be off by a quintillion or two, but who's counting?

CHAPTER 8

Man-Made Lands

Josef Kleindienst, tall, urbane, and self-assured in a cream-colored suit and cornflower-blue shirt, was pleased to welcome me to his own personal Germany. I followed him off his sleek little yacht and onto a beach marked with a wooden sign striped with the colors of the German flag. WILLKOMMEN IN DEUTSCHLAND, it read.

At the time of this visit, in late 2015, it was difficult to perceive the Teutonic character of the place. First off, the weather was warm and sunny, though Christmas was just weeks away. Second, it's an island. Actually, it's not even a real island. It's an enormous mass of sand dredged from the bottom of the Persian Gulf and piled up a couple of miles off the coast of Dubai, one of the seven micro-kingdoms that make up the oil-rich United Arab Emirates. A bunch of small olive and palm trees sat in pots near a little gazebo sheltering a few golf carts. Two workmen in yellow safety vests and helmets loitered around, while a third walked along the waterline looking for nonexistent trash. Other than that, Germany was just fourteen acres of flat, barren sand.

But Kleindienst, an Austrian-born real estate developer, has a vision of something far grander. Kleindienst has spent tens of millions of dollars to make this pile of sand, and five others connected to it by little bridges, into a luxury resort simulacrum of his home continent that he believes will prove irresistible to legions of holidaymakers and vacation-home buyers from around the world. Each of the six islands is based on a different country or region—Germany, Monaco, Sweden, Switzerland, St. Petersburg, and "Main Europe." Sweden will feature sauna-equipped private villas roofed with the inverted hulls of Viking ships. Monaco will host a "seven star" hotel and marina complex, featuring elements from the life of the late Princess Grace Kelly. The heart-shaped St. Petersburg will host classical ballet and opera performances. The vaguely defined island of "Main Europe" will be highlighted with a pseudo-Viennese city street with piped-in rain. In a final cherry of hubris on top of this titanic chutzpah sundae, Switzerland will feature a faux city street where real snow, blown from artfully concealed rooftop pipes, will drift down on strolling tourists.

"On one side you will have snow every day, and other side, a tropical beach," enthused Kleindienst in his Schwarzeneggerian accent. "We will create a space that is not existing anywhere."

Kleindienst took the wheel of one of the golf carts to show me around the naked little cays that will soon emerge as central Europe. "What we bought [in 2007] was just a pile of sand," he said. In 2015, it still wasn't much more than that. The only island with any action was Sweden, which is to be the most exclusive of Kleindienst's islands (though why he believed a Nordic country that's cold and dark half of the year will be a big draw in the Middle East wasn't clear). A couple of dozen laborers and a small collection of bulldozers, front-end loaders, and trucks bustled around a handful of pocket-size construction sites. Mostly they were just deep holes, with puddles of seeped-in seawater on the bottom. They were to be sealed with concrete and made into foundations for the islands' ten private villas. The 20,000-square-foot pleasure palaces will include seven bedrooms, a sauna, a "snow room," a home theater, and a gym; for an extra fee you can also have the villa furnished by Bentley, the luxury car company. Asking price: about $13 million. The villas will come equipped with elevators, which is important because they will be five floors high, each with a private disco on top. "That might get a little loud," I said. "We hope so! It's a party place," enthused Kleindienst.

Ultimately, said Kleindienst, the whole project, dubbed the Heart of Europe, will encompass 4,000 housing units, twelve hotels (including the only one in the UAE where you can bring your dog), and dozens of restaurants. There are no roads to the islands, nor any on them; visitors will have to arrive by boat, helicopter, or seaplane.

I lived in Las Vegas for a while, which made a lot of this seem familiar. Kleindienst's project reminded me of one of my favorite of that city's epically scaled theme hotel/casinos: the Venetian, with its indoor gondola canals and ersatz St. Mark's Square.

Kleindienst did not appreciate the comparison.

"It's not like the Venetian," he said disdainfully. "They're coming from a theme park angle. We are building a leisure destination with elements of different countries. Each restaurant will be staffed by people from that country, and they'll behave like they do in their own country. We want to offer an authentic experience." The place will even use the euro as currency instead of the UAE's dirhams. Those trees now scattered around Germany's landing dock are centuries-old olive trees imported from Spain; they'll be deployed to give Monaco a genuinely Mediterranean feel. There will be street performers, artists, musicians, even a circus, all brought in from Europe.

"Europe has fifty-one countries," Kleindienst continued. "Every week we'll have a festival from one. We'll have a typical restaurant from every country, artists from every country. We want the Finnish restaurant to welcome people in Finnish. You should be able to experience Europe here in the Heart of Europe."

If you've got the money to travel to Dubai but you want to experience Europe, I wondered, wouldn't you just, um, go to Europe? "Sure, you can go to the real country," replied Kleindienst, who had clearly been asked this before. "But you can't go to the beach in Finland year-round. Here you'll have the chance to experience food from fifty-one countries, festivals, street artists from fifty-one countries, everything in one place." Not to mention underwater villas, fireworks shows, and snorkeling facilities.

Kleindienst said he was aiming for a grand opening in 2020. He had already pushed back the finish date more than once, however. At the time of this writing, in late 2017, construction was still grinding forward.

As Brobdingnagian as Kleindienst's plans are, they are only a tiny part of a much, much bigger project. Kleindienst's six island mini-nations make up just one neighborhood of the World: an archipelago of some three hundred artificial islands, roughly forming a map of, well, the world, built at the behest of the Emir of Dubai. Hundreds of millions of tons of sand were dredged from the Persian Gulf to form them in a paroxysm of speculation-driven geo-engineering in the mid-aughts. It is likely the biggest assemblage of artificial land ever created.[1]

The idea was that developers and the global 1 percent would buy the islands and convert them into their own whimsical versions of the nations they represent. But when the global recession hit in 2008, the World came to a standstill. When I visited in 2015, almost every one of these hundreds of "islands" was still just a low, flat hummock of bare sand leopard-spotting the Persian Gulf's surface, like blobs of cookie dough on a big blue tray.

The whole project is, of course, ridiculous. But Dubai, once a tiny fishing village that is now home to the world's tallest building, the world's biggest shopping mall, and an indoor ski hill, has proven many times that just because something is ridiculous doesn't mean it's a bad business idea. And that definitely includes enormous, fancifully shaped artificial islands. Dubai is also home to the Palm Jumeirah, a man-made peninsula in the shape of a palm tree, a landmass so big you can see it from outer space. It hosts a gobsmacking panoply of luxury apartments, villas, and resorts where tens of thousands of people work, live, and play. In other words, where there was only water fifteen years ago, Dubai has created billions of dollars worth of real estate out of plain old sand.

The Palm Jumeirah and the World are only the most ostentatious of many such "land reclamation" projects around the gulf and around the world. From the South China Sea to Tokyo Bay, from California to Nigeria, humans are putting unprecedented volumes of sand to one of its most consequential uses: the godlike power to create new land. We have disinterred vast armies of construction and silica sands from the ground and put them to work in ways that have transformed how we live; now we are also dredging enormous legions of marine sands from the ocean floor and using them to literally change the world, to alter the shape of countries and coastlines and create new land where there was no land before.

Deployed in this way, marine sands are converted into valuable real estate. In some places, they are also made into a tool of geopolitics, a weapon with which nations assert themselves at the expense of their neighbors.

uy land," Mark Twain once famously said. "They're not making it any more." Clever quip, but completely wrong. The Dutch have been building artificial land, much of it below sea level, since the eleventh century, damming wetlands and pumping them dry.[2] Peter Stuyvesant, the first governor of what would later be called Manhattan, began expanding the island back in 1646, mostly with earth displaced by the construction of buildings and canals. Sand, however, is the material used most often to create new land. In the 1850s, developers filled shallow areas of San Francisco Bay with sand scraped from nearby hilltops to create what is now the city's financial district.[3] A decade later, the introduction of centrifugal pumps, powered first by steam, then diesel, enormously increased the amounts of sediment that could be siphoned from the ground under the sea. They also make it possible to pump the stuff through miles of pipeline to distant destinations. Sand dredged from underwater built long stretches of Chicago's lakefront,[4] as well as large portions of Marseilles, Hong Kong, and Mumbai. Elsewhere in the United States, sand has been used to create artificial islands from scratch, including San Francisco's Treasure Island, Southern California's Balboa Island, and Seattle's Harbor Island.

But those efforts are puny compared to the gargantuan scale and audacity of modern land reclamation projects. What's driving those efforts is a familiar force: the ever-swelling movement of people into cities.

Because cities require trade to thrive, they tend to be sited on lakeshores, rivers, and especially seacoasts. Cities are attracting millions more people every year, and port cities are some of the most attractive: Eight of the world's ten biggest cities are on the ocean. Fully half of the world's population lives within sixty-two miles of a coastline.[5] Those cities need space to house all those people, not to mention for the factories, ports, and other places where those people work. Many seaside megacities, from Tokyo to Lagos, are already densely packed, but are hemmed in by mountains, rivers, or deserts, making it tough to expand farther inland.

Sand, it turns out, can not only make the concrete and glass for the buildings sheltering those people, but also the ground on which those buildings sit. Beginning in the 1970s,[6] advancing technology made it easier and cheaper to simply create more land. Bigger dredging ships equipped with extremely powerful pumps came on the market, capable of hauling up marine sand from ever greater depths and delivering it in ever greater quantities with ever greater accuracy onto predetermined places. In 1965, the largest dredges could hold about 6,500 cubic yards of material. That number more than tripled by 1994; by now it has grown nearly tenfold. As of 2017, the biggest dredge in operation was more than 700 feet long; stood on end, it would overtop a sixty-story apartment building. It carries a pipe that can pull up sand from 500 feet below the water's surface.

Land reclamation generally requires sand similar to that used in concrete: angular, interlocking, medium-sized quartz grains. According to the International Association of Dredging Companies, if good quality sand is available within a reasonable distance, new seafront land can be built for less than $536 per square meter—a fraction of the cost of buying existing seafront land in hot spots like Hong Kong, Singapore, or Dubai.[7]

The new generation of dredging techniques was first applied on a major scale in the early 1970s to expand the Dutch port of Rotterdam into the North Sea. In 1975, Singapore followed up by building a new airport on top of 40 million cubic meters of sand pulled from the seabed. (Airports in Australia, Japan, Hong Kong, and Qatar have also since been built on reclaimed land.)[8] In the years that followed, land for industrial estates was created in Tokyo Bay with sand from over 240 feet below sea level. More deep suction dredging built up the coasts of Singapore, Taiwan, Hong Kong, and Amsterdam. China, the fourth-largest nation on Earth in terms of naturally occurring land, has added hundreds of miles to its coast, and built entire islands to host luxury resorts.[9] Lagos, Nigeria, is adding a 2,400-acre urban extension to its Atlantic shoreline. Smaller nations, including the Maldives, Malaysia, and Panama, have also built islands from scratch.

Dubai's neighbors have also decided they could use some more oceanfront. Qatar has built nearly 1,000 acres of land out of sand just off the coast of its capital, Doha. Bahrain has built up its harbor and a set of from-scratch resort islands out of dredged sand held in place with enormous tubes filled with more sand.

And then there's Singapore, a world leader in land reclamation. It is one of the most densely populated countries in the world, extremely rich but geographically tiny. To create more space for its nearly 6 million residents, the jam-packed city-state has built out its territory with an additional fifty square miles of land over the past forty years, almost all of it with sand imported from other countries. As I mentioned in chapter 1, the collateral environmental damage has been so extreme that neighboring Indonesia, Malaysia, Vietnam, and Cambodia have all restricted exports of sand to Singapore. Nonetheless, according to local media and outside organizations, sand still flows illegally from those places, especially Cambodia.[10] Singapore has also cast its net farther afield. It now buys sand from Myanmar, Bangladesh, and the Philippines. The country is so anxious about supplies that it stockpiles a strategic reserve of sand for emergencies.[11]

All told, according to Deltares, a Dutch research group, human beings since 1985 appear to have added as much as 5,237 square miles of artificial land to the world's coasts—an area about as big as Connecticut or the nation of Jamaica.[12] Much of it with sand.

Even amid such competition, Dubai's artificial islands take the prize for sheer brazenness. The International Association of Dredging Companies calls them "the most ambitious reclamation projects of all time in terms of size, concept, and engineering."[13]

That Dubai would create such vast, record-setting construction projects is all the more astonishing when you realize just how tiny and insignificant a place it was until very recently.

Arab herders and desert-dwelling nomads have lived in this desolate corner of the Arabian Peninsula for millennia. The environment is so harsh that the population stayed stuck at around 80,000 from the advent of Islam in AD 630 until the 1930s.[14] As former journalist Jim Krane writes in *City of Gold*: "Those who eked out a living [in what is now the UAE] were, until about fifty years ago, among the planet's most undeveloped societies. No one envied their existence of perpetual hunger and thirst, nor their diet of dates and camel's milk."[15]

One of the few permanent settlements was a tiny fishing village called Dubai that had grown up along an inlet from the Persian Gulf. Local tribes tussled over the meager glory of ruling the place until imperial Britain, concerned mainly with protecting its sea routes to India, began to assert itself in Arabia in the early 1800s. Dubai's ruling sheikhs kept their positions and a measure of independence by cutting peace treaties with it. Great Britain never formally colonized the area, but dominated the sheikhdoms that would become the UAE for the next 150 years.

In 1833, Dubai was overrun by tribespeople led by one Sheikh Maktoum bin Buti. The British soon recognized the Maktoums as

Dubai's rulers, a position the family has held ever since. Power has passed peacefully from one sheikh to the next for 175 years—by Middle Eastern standards, an extraordinary streak of stability.

In those pre-oil days, pearls were the gulf's top commodity. Divers jumping from boats hauled up the precious baubles by the sackful, making local merchants rich. But the market collapsed in the depression of the 1930s, and what was left of it was taken over by the invention of cheaper "cultured" pearls in Japan. In Dubai, businesses went bankrupt, merchants left town, and the local economy was hit so hard that food became scarce. "As World War II ground on," writes Krane, "the famine grew desperate. When there was no rice, fish, or dates, people ate leaves or the ubiquitous dhub, a spiny lizard that may have given Dubai its name. Plagues of locusts became a blessing. People would net the bugs and fry them, crunching on them by the handful. . . . Inevitably, some Dubaians starved to death."[16]

Things improved after the war, but Dubai was still a backwater barely noticed by the outside world. In the 1950s, many of its 15,000 inhabitants still lived in palm-thatched *barasti* shacks and adobe houses, and camels wandered the sandy pathways of the town. Electricity and manufactured ice—of perhaps equal importance to the locals in a place where temperatures routinely hit triple digits—arrived only in the early 1960s. But things were about to change, and change radically, because in 1958 oil was discovered in neighboring Abu Dhabi.

Abu Dhabi turns out to have gargantuan amounts of oil—at least 92 billion barrels, worth trillions of dollars. Dubai found respectable amounts of offshore oil starting in the late 1960s, but nothing close to Abu Dhabi's winning geological lottery ticket. Today Abu Dhabi produces some 2.5 million barrels of oil per day;

Dubai barely manages 60,000. In fact, the emirate is now a net importer of oil and gas.

What Dubai lacks in fossil fuels, though, it makes up for in a commodity far more rare in the Middle East: competent leadership. The Maktoums began building up their picayune port into a hub for business and trade over a century ago. In the early 1900s they abolished customs duty and lured Arab and Persian merchants with offers of free land and promises they could do business unhindered by government (which also encouraged the growth of a lively smuggling trade of everything from drugs to gold that still endures). Streams of fortune-seeking immigrants, especially from what is now Iran, moved in, swelling the population and its trading links abroad. Today there are nearly three times as many Iranians living in Dubai as native Emiratis.

In 1958, the emirate's fortunes took a decisive turn thanks to an early bit of land reclamation. The inlet, known as Dubai Creek, had been silting up for years, forcing incoming ships to anchor offshore. (Sometimes sand just gets in the way.) Sheikh Rashid bin Saeed al-Maktoum, father of today's ruling Sheikh Mohammed bin Rashid al-Maktoum, dredged the creek deeply so that it could take in more and bigger ships. Well aware of the value of sand, the sheikh then used the dredged grains to build up land on the creeks' banks, which he sold to merchants. It was a double win.

"From then on," writes Krane, "Dubai would ride an incredible growth spurt that has yet to stop. The dredging of the creek was the spark that started the whole thing."[17] The sheikh sank oil profits into building ever more expansive roads, ports, airports, and state-owned businesses, including Jebel Ali, the world's largest man-made harbor. It may have still been a tiny statelet, but Dubai made its grandiose ambitions plain. "In 1974, they built the World

Trade Center, with a Hilton hotel in it. It was then the tallest building in the Middle East—in the middle of nowhere!" chuckles George Katodrytis, a professor of architecture at the American University of Sharjah, the emirate next door to Dubai.

Money and people poured in. In 1960, the city had 60,000 residents, most of them living in an area of two square miles. Twenty years later, it had ballooned to 276,000 people living in thirty-two square miles.

All this development has certainly worked out well for the ruling family. Sheikh Mohammed is one of the richest men on Earth, with a fortune estimated in the double-digit billions. Dubai was already booming when he pushed it into overdrive in 2002 with an unprecedented move: Dubai, he decreed, would allow foreigners to buy homes. This was something no other gulf country allowed. It turned out to be a masterstroke that triggered a real estate boom of global proportions—one that soon led to the need to create more real estate, in the form of islands.

Dubai is tremendously appealing to a certain type of global citizen. It has an excellent banking system with Swiss-like opacity. It imposes no taxes and few restrictions on imports and exports. It boasts good schools, hospitals, and infrastructure.

Above all, Dubai offers safety. It is a literal oasis of security and political stability in the world's most chaotic region. It offers a haven for anyone in Iraq, Pakistan, Libya, or any other nearby country who fears that war, economic chaos, or the attention of government officials might threaten their business. It's a safe place where such people can base their companies and park their money and even their families.

Dubai's wealth helps keep it stable, along with its ironfisted intolerance for political dissent. According to Human Rights Watch, "The government arbitrarily detains, and in some cases forcibly disappears, individuals who criticized the authorities, and its security forces face allegations of torturing detainees." It's an effective combination. Dubai has never suffered a coup or civil war, and hasn't had a significant terrorist attack in over fifty years.

It is also uniquely open and tolerant by gulf standards. Though the native Emiratis tend to be conservative Muslims, recognizable in their spotless traditional white robes and head coverings (Sheikh Mohammed is himself a husband to several wives and father to at least twenty-four children), everyone else is more or less free to do as they like. There are plenty of Hindus, Christians, even Jews living and working there.

At the same time, Dubai has improbably made itself into a popular tourist destination. Sure, it's located in one of most chaotic and repressive regions of the world. But it's also sunny almost all year round, with great beaches and a warm sea. And unlike neighboring Saudi Arabia and Kuwait, you can drink alcohol, dance at a nightclub, and sunbathe in a bikini all you want in Dubai. The emirate had only forty-two hotels in 1985. Today it has hundreds, hosting more than seven million visitors each year.

The whole enterprise got an unexpected boost thanks to the 9/11 terror attacks, of all things. In its scramble to shut down terrorist financing networks, the United States froze the bank accounts of some gulf Arabs they thought might have links to Al Qaeda. America was suffused with suspicion toward Middle Eastern money. Many wealthy Arabs and their money managers decided they'd be better off keeping their wealth closer to home. And so billions of dollars flowed out of America and headed east, looking for a safe harbor to invest in. Dubai stood there smiling, holding its doors wide open, and the money rushed in.

The result was that Dubai's real estate market exploded. Office towers, shopping malls, and luxury hotels burst up from the ground.

The city that has ensued—and which is still growing at a baf-fling pace—is easily one of the weirdest places I've ever been. It's a fantasy conjured up like a genie out of the desert. Dubai represents the triumph of the power of money and will over nature; how else to explain the presence of not one but many golf courses and orna-mental lakes in the middle of the desert? Of gigantic islands built of sand where there was once only water?

Riding the $8 billion state-of-the-art driverless Metro along its elevated track from one end of the city is to enjoy an eye-widening tour through the heart of a futuropolis straight from some Pixar sci-fi fantasy. It's a miles-long belt of urban density, packed with high-sheen glass-faced skyscrapers and asphalt roadways eight lanes wide, all of it sandwiched between swaths of sand—the desert on one side and the beach on the other. There are buildings fifty stories and more in all manner of fanciful shapes—one twisted like a cork-screw, one shaped like a half-moon, another a set of concentric half-circles. Towering far above them all is the surreal spire of the Burj Khalifa, the tallest building in the world, surrounded by a dense thicket of towers so emphatically dwarfed that they seem to be gazing up in awe at their gleaming, glass-faced leader.

At the city's margins, swooping elevated highways with clover-leaf interchanges lace the desert. And in seemingly every unoccu-pied space, cranes sprout up like giant steel dandelions, lording over construction sites crawling with bulldozers and backhoes and workmen in yellow safety vests.

Think about how much sand it took to create such a city out of practically nothing, most of it in the last twenty years. That's why sand is now starting to become a serious issue. We have never con-sumed this resource at a pace remotely like the quantities we are consuming now.

Dubai's overall vibe is sort of like a gigantic open-air airport lounge. It's clean and modern, full of concrete and glass buildings housing familiar chain stores and fast-food restaurants and ads for famous brands. On the street, in the malls and hotel lobbies, you could be in any twenty-first-century city, any place that attracts people from around the world. It's a kind of postmodern city, a place stripped of any identity besides modernity itself.

The whole city seems strikingly out of place in a region so steeped in history, religion, tradition, and culture, where adherence to ancient faiths and traditions is so prized. But it has proven a phenomenally successful model. Dubai is the Middle East's leading financial hub, home to its biggest port and to the world's busiest and biggest airport. The five-star hotels keep booking up and the luxury villas keep getting bought.

Dubai is selling, above all, location—a desirable physical place. But once it really started booming, it ran into a problem: it doesn't actually have much space. At least, not the most desirable kind to tourists and well-heeled home buyers, which is of course beachfront property. Dubai aims to pull in 20 million tourists by 2020. The emirate has only forty miles of natural coastline, and it was getting built up fast. The solution was as obvious as it was implausible: build more.

When the island-building campaign got going in the mid-1990s, the original plan was to build a conventional-looking round island. But that would have added only a few miles of beachfront. Sheikh Mohammed, or so the official story goes, came up with the idea of a design that would both evoke the emirate's culture and also provide far more beach: a palm tree, each frond of which would be a spit of beach-edged real estate. The Palm Jumeirah would be the first piece of artificial land ever built deliberately designed to be a shape you could identify from the air. It would more than double the emirate's coastline, adding forty-eight miles of new shore, including thirty-eight miles of beach.

Sheikh Mohammed created a new state-owned company, dubbed Nakheel, meaning palms in Arabic, to build the island. He put a trusted lieutenant, Sultan Ahmed bin Sulayem, in charge. To build the land, Nakheel turned to Van Oord, a venerable Dutch company that is one of the biggest dredging and land reclamation outfits in the world. (After all, the Dutch have centuries of expertise in the trade.)

Dubai sits right at the edge of one of the world's great sandpiles—the vast desert of the Arabian Peninsula's Rub' al Khali, or Empty Quarter. But desert sand doesn't work for land reclamation any better than it does for concrete: the grains are too rounded to lock strongly together. Luckily, there is plenty of usable sand on the other side of Dubai. The only complication is that it's at the bottom of the Persian Gulf.

That was no problem for Van Oord. The company sent out self-powered, self-guided surveying ships to take core samples from the nearby seabed, looking for sand that had the right chemical composition, amount of organic material, and compressive strength for the job. Once the scouting craft located a deposit of sand appropriate to the job, six miles offshore,[18] a fleet of dredging ships, guided by GPS, went out to what are known as borrow areas. The ships lowered enormous pipes, fitted with screens to keep out anything bigger than a fist, into the water, and vacuumed up the sand into their holds.

Then each ship sailed back shoreward to where the GPS told them the Palm was to be, opened its hold doors, and simply dropped the sand. After a few rounds of this, the pile would get too big for the ships to clear. At that point, the ships would stop a few hundred yards away, tilt skyward a hose the size of a cannon, and shoot a torrent of slurry into the air. The process is called rainbowing, which definitely sounds prettier than "blasting five tons per second of sand and water through the air." Again guided by GPS, the movable nozzles drew the shape of the Palm like gigantic sand-shooting spray paint cans.

In some ways, the gulf is a uniquely well-suited platform for such a project. It's shallow, reaching only 300 feet at its deepest, which makes it relatively easy to pile up enough sand to rise above the waterline. It's also sheltered and relatively calm, with few waves to erode the sand piles.

Still, as anyone who has ever built a sand castle knows, just piling up grains makes for a pretty loose structure. That's not good when you're planning to put thousands of tons of buildings on top of it. To firm up the new land, Van Oord applied vibrocompaction, which involves cranes driving fat metal spears deep into the sand and setting them vibrating. The vibrations make the grains jump and shake around, settling into the void spaces between them, locking them together, making the structure denser and more solid. It also makes the pile smaller, so more sand has to be added again after the vibrocompacting.[19] The process involved drilling more than 200,000 holes over a period of eight months.

When the Palm Jumeira was finally completed in 2005, some 120 million cubic meters of sand had been piled up to form it. The whole island is surrounded with a breakwater built of rocks piled atop yet more sand. All told, there's enough sand and rock in the Palm to circle the globe with a wall seven feet high.

Here's an equally amazing fact about the Palm Jumeirah: Nakheel sold every lot on it before they'd even built the place. In May 2001, while there was still nothing but water where the Palm was to be built, Nakheel declared it open for business. Up for grabs were

2,500 beach apartments along the trunk, and 2,500 private villas on the fronds. The villas started at $1.2 million US for a four-bedroom "Garden" villa, going up to six-bedroom "Signature" villas; all came with two parking spaces and a maid's room, of course, and a patch of sandy beach right outside the back door. At least that's what the architect's drawings showed; there was nothing actually built yet. "When a potential buyer turned up, bin Sulayem, an elegant man with a perfectly clipped mustache and an ingrained politeness, fired up his speedboat," writes Krane. "He'd zoom investors a mile and a half out to sea . . . Then he'd cut the motor. 'This is where your villa's going to be,' bin Sulayem would tell his client, as they bobbed in the Gulf. 'Now give me a deposit.'"[20] Every single one of them was snapped up within seventy-two hours.

It was only a few years later that construction of those villas actually got under way. Forty thousand laborers were put to work deploying millions more tons of sand, in the form of glass and concrete, into place.

The buyers—some private, some resellers—came from some thirty different countries. About a third were gulf nationals and a quarter Brits (including David Beckham), and there was at least one Austrian—Josef Kleindienst.

Kleindienst first came to Dubai in 2002, his money-seeking antennae having picked up the signal sent by the opening of Dubai's real estate market to foreigners. He had been a Viennese police officer for eighteen years, and claims to have enjoyed it, rising to the post of inspector general. He also dabbled in politics, joining Austria's far-right Freedom Party and heading a police union affiliated with it. He broke spectacularly with the party in 2000, publishing a book titled *I Confess,* in which he accused party leaders of paying police to illegally slip them classified information, a charge the party denied.

All the while, he had a sideline in real estate. He'd watched his father and grandfather buy and sell land since he was a child. Kleindienst saw the door of opportunity swing wide open in the early 1990s, when Communism collapsed in Europe and all of Austria's eastern neighbors were suddenly open for business. Working with friends in the Hungarian police, Kleindienst snapped up a bunch of plots in Budapest. "We made very good money," he said. By 1999, he had quit the police force to go into real estate full-time, launching the company now known as the Kleindienst Group. The company now has investments in properties across central Europe as well as in Pakistan, Seychelles, and South Africa. But it was in Dubai that Kleindienst really found a canvas adequate for his ambitions. He started operations there in 2003, and since then his company has developed a range of apartment complexes, office parks, and hotels in the emirate.

"We bought fifty villas on the Palm Jumeirah," he says. "It was only sand then. We would have bought more, but there was nothing else available! We came a couple of days too late."

The Palm Jumeirah was such a hit that plans were quickly announced to follow it up with two more palm-shaped islands that would be even bigger—the Palm Jebel Ali and Palm Deira. By then, so much sand had been dredged from the gulf's floor that the quality of the remaining grains was declining, requiring additional time and expense for extra vibrocompaction.[21] No matter. In the frothy financial climate of the mid-aughts, there seemed to be no limit to how much new land could be built and sold.

So Nakheel started work on the most audacious project yet: the World. Never mind buying villas; now you'd be able to buy a whole country.

Construction began in 2003. The World would create over 2,500 acres of new land out of 320 million cubic meters of sand, adding 144 miles to Dubai's coastline. Planners expected the islands to host as many as 300,000 people. Unlike the Palm, where Nakheel built up many of the buildings and infrastructure, the World islands were to be sold empty, tabulae rasae for developers' dreams. Investment costs: an estimated $14 billion.

"The World was millions of dollars worth of branded sand," says Adnan Dawood. He should know. For several years while the World was under construction, Dawood was in charge of marketing and public relations—the man tasked with selling the world on the idea of the World. In 2003, he had just graduated with a marketing degree from California State University, Fullerton, and had taken a humdrum job with a local tile company. Dubai was just getting started on its most grandiose projects, and Dawood figured it would be a more exciting scene to be part of than selling floor tile in Southern California. He convinced Nakheel to hire him and rode the wave upward.

Dawood is a trim American-educated Muslim of Indian descent in his late thirties whose family came to Dubai when he was two years old. He left Nakheel in 2009, but he loves talking about his time there—the big money, celebrities, glamour.

Dawood started at Nakheel in 2005, while the sand was still being dredged. The islands, measuring between just under three acres to more than ten acres, were priced from $15 million to $50 million. Sales were going slowly. "We had a logo, we knew what we wanted to call it, but that was it," Dawood said. "There was no strategy." Dawood decided to take aim at a fat target: the vanity of the rich and powerful. "We started telling people they couldn't buy in," he said. Nakheel claimed that each year only fifty individuals, chosen on the basis of their 'achievement,' were invited to buy an island. "This was in 2006, 2007, when ego and cash were a deadly combination. The moment we said, 'You can't have it,' everyone wanted it."

Dawood and his team touted the islands as the ultimate luxury home for the one-percenter who already has everything, a private Bond-villain island nation. They played up the illusion of elite exclusivity by piling on glamour. They coaxed Annie Leibovitz into doing a photo shoot with Roger Federer on the islands. They gave well-publicized tours to celebrities from Michael Jackson to Malcolm Gladwell to Donna Karan and Donald Trump's son Eric. A cunning look creeps into Dawood's wide-set eyes as he details his tactics. The press loved speculating about who might buy which "countries," especially the United Kingdom. So when Dawood learned publicity-happy billionaire Richard Branson was coming to Dubai to promote his airline, he proposed a twofer. Dawood rounded up a bunch of foreign journalists and took them on a boat out to the World. As they drew near the top of Europe, everyone noticed a classic English phone booth sitting in the sand at the water's edge. Suddenly out popped Branson, dressed in a Union Jack–striped suit and waving a British flag overhead for good measure. The picture got both the World and Virgin a torrent of giddy publicity. "The funniest thing is, he was actually on Denmark," said Dawood. "The UK island wasn't even built yet!"

The celebrity rumor-mongering was such an effective attention-getter that at one point Dawood leaked word to a local news website that Brad Pitt and Angelina Jolie were buying Ethiopia. Then he called some bigger publications to make sure they saw the website's anonymously sourced article. Within days, outlets from CNN to *People* magazine were breathlessly reporting on the couple's latest excursion to "Africa." In the end, it didn't really matter to anyone that the story was completely made up.

What did matter to Nakheel was that the islands started selling. By 2007, some 70 percent of the islands had been sold. Lauren MacDonald, a Canadian marketing executive, was working in London for Pepsico in 2008 when she got what sounded like an amazing job offer. Nakheel wanted her to come work on the World. They offered to triple her salary, which would also be tax-free in Dubai. It seemed like a no-brainer.

But then came the crash in 2008. "I signed with Nakheel three weeks before Lehman Brothers collapsed," she said. "I kept calling Nakheel, saying 'It seems the situation is terrible!' They said, 'No problem.' Well, within three months, that company went from four thousand employees to six hundred. I worked there for a year and a half, and the whole time I didn't know if I'd have a job the next day."

She did manage to sell two islands, each for tens of millions of dollars, she said—Taiwan to an Italian hotelier and Iceland to a German. "They were über-rich individuals who felt it was time to buy because you could get bargain-basement prices."

Just as the World had been getting in gear, the river of cash that had birthed it suddenly dried up. Out in the actual world, the financial crash of 2008 wiped out billions of dollars of investor capital. A pile of naked sand in the middle of the Persian Gulf suddenly didn't seem like such a great investment after all. Dubai's entire real estate market tanked in spectacular fashion. At one point, the emirate was so broke it had to borrow $10 billion from its wealthy neighbor, Abu Dhabi.

Nakheel laid off hundreds of employees, including Dawood. Several executives were arrested in ensuing corruption probes, resulting in prison sentences for at least two of them.

The artificial island building basically ground to a halt. All the sand had been put in place for the Palm Jebel Ali, but that was as far as the project got. It remained an empty, artfully shaped pile of grains, unblemished by a single road or structure. The proposed Palm Deira was put on hold. And the World stopped.

Developers went broke. Some went to jail for bouncing millions of dollars in checks. Investors filed lawsuits over hotels and villas that never got built. At least one committed suicide.[22] Kleindienst's company nearly went bankrupt. "It was very stressful. I had to lay off people who were good friends. I had to tell them, 'Guys, you have to find another job,' but they knew there was no other job to go to," Kleindienst told Britain's *Daily Mail* in 2010. "For many people the Dubai dream was over."

For people concerned about the natural environment, though, it was more like a respite from a nightmare. Building new land in the water may be good business, but it is brutal on the ecosystem. "Land reclamation is one of the top three causes of damage to the Persian Gulf," said John Burt, a marine biologist at New York University's Abu Dhabi campus who has been studying the gulf ecosystem for years.

For starters, pulling those huge armadas of sand up off the ocean floor destroys the habitat of whatever was living there. "Engineers call them 'borrow areas,' though they never return what they're borrowing," said Burt. "They have a lot of great euphemisms for environmental degradation." The borrow areas are typically just sandy bottoms with little evident life. Still, says Burt, "I'm sure there are organisms there that just haven't been documented."

All the sediment that gets stirred up by these dredging operations also clouds up the surrounding water for what can be a long time. It's the same issue that bedevils beach renourishment projects in Florida, only on an even larger scale. The increased turbidity— the amount of sand and silt suspended in the water—can essentially suffocate fish, crustaceans, and other creatures. It also blocks sunlight from reaching plants deep below the surface.[23] That's not just a concern for academics like Burt: an island-building project in eastern Indonesia was put on hold in early 2017 after a series of protests by local fishers who feared the dredging would wipe out local fish stocks.

Then there's the issue of what all that sand gets put on top of. The Palm Jumeirah was built on a flat, sandy bottom. The millions of tons of sand that made the Palm Jebel Ali, however, were dumped right on top of three square miles of coral reef.[24] The reef had been designated a protected area, but such considerations tend to take a back seat to development in the gulf. In nearby Bahrain, a far bigger coral reef has been almost completely destroyed, largely thanks to land reclamation.[25] Other projects in the gulf have buried oyster and sea grass beds.[26] China has constructed huge tracts of land along its coast in recent years, wiping out wetlands and shorebird habitats.[27]

The man-made landmasses also shift the pattern of the gulf's currents such that they no longer carry sand to existing beaches. As a result, Dubai has had to spend millions in recent years to replenish some of its mainland beaches with sand trucked in from construction sites.

"Humans will continue developing coastlines, but there are ways to do it more sustainably," said Burt. "Within one or two generations we'll have lost most of the ecosystems along the coast. I'm not a citizen of the UAE. But if I were, I'd be pretty upset about what I was leaving behind for my kids and grandkids."

Brendan Jack, a Nakheel spokesperson, was quick to assure me about the company's concern for the environment. Nakheel transplanted some of the coral from the reef now buried under the Palm Jebel Ali, used independent consultants to find areas where it could extract sand with minimal damage, and made Van Oord set up underwater curtains to limit the spread of silt during dredging, he told me. And as a bonus, the rocky breakwaters around the islands are now home to all kinds of marine life.

"True, there are impacts, and it's not the same as before," he said. "But that's true of any construction activity. That's just reflective of human activity anywhere on the planet. It's all pros and cons. We try to minimize the impacts and maximize the benefits."

It's Jack's job to explain or excuse the actions of his employer, but he does have a point. Any kind of development entails environmental costs. The world can take the loss of a coral reef here or a fish habitat there, tragic though they may be. But focusing only on such local impacts is missing the forest for the trees. The bigger question is, can the planet handle the whole way of life that Dubai both represents and embodies—the air-conditioned, car-dependent, energy-guzzling, resource-intensive "good life"? With that in mind, it's worth knowing that residents of the United Arab Emirates lead the world in per capita consumption of water and electricity, and in waste production. These desert dwellers use 145 gallons of water per person per day, the highest rate in the world.[28]

Meanwhile, the global economic recovery set the money train rolling again. Kleindienst's investors returned, and construction started up again in 2013. The other islands are also showing signs of life. The Palm Jebel Ali was still on hold in late 2015, but building was under way on the Palm Deira, now recast as Deira Islands, a smaller, more conventional-looking development including a marina, hotels, apartments, and acres of malls. In early 2017, another group of investors announced plans to build two new artificial islands, adding another 1.4 miles to Dubai's coast.

On the yacht ride back from the World, we relaxed on the upper deck on buttery white leather couches, enjoying the sunshine and warm breeze. A demure young woman in a white uniform poured us Moët in flute glasses. Bowls of fruit, nuts, olives, and, inexplicably, Doritos were scattered around. (I munched on a few of the chips, because when would be the next time I'd have a chance to pair Doritos with champagne?)

I had a last question for Kleindienst. "In a lot of the international press, even among people here in Dubai, people laugh at the idea of the World," I said. "They say it's a failure, a big joke. Does that bother you? Do you think about that?"

Kleindienst paused for a moment, then responded with a short lesson about modern Dubai's short history. "The father of Sheikh Mohammed, Sheikh Rashid, he decided to build Jebel Ali port," he said. "When he decided to build this port, his people asked him if he's crazy. Why he is building out there this huge port? Today it's one of the biggest and busiest ports in the world. Attached to it is a free zone with more than two thousand companies. And this is a cornerstone of Dubai's success. Nobody is laughing about him today anymore. And when they see [the Heart of Europe] built, nobody will laugh anymore."

The next day, I went to see the Palm Jumeirah, then the only completed version of all Dubai's artificial waterborne Edens. It has continued to grow ever more elaborate and lavish. After a look at some of the more opulent hotels, I visited the home of Carrie Hart, a slender, elegant entrepreneur, festival promoter, and devoted Burning Man attendee originally from Minnesota. She moved to Dubai a few years ago with her oil trader husband and their two small children.

We sipped mint tea and snacked on delicate little pastries on the poolside verandah of her marble-floored villa on Frond F of the Palm Jumeirah. A few yards away, the golden sands of the artificial beach sloped gently away into the turquoise waters of the gulf. A few hundred yards across the water lay Frond E, edged by a similar beach and lined with a similarly lush collection of homes. Behind them, back on the mainland, loomed the skyscrapers of Dubai.

Hart and her husband were attracted here from London, where they had been living, by the warm weather, safe streets, and generally high quality of life. "We decided to live on the Palm because why live in the desert when you can live by the sea?" she said. Her kids go to a private school and play ice hockey in the Dubai Mall's indoor rink. It's very safe; each frond has a gate monitored by a guard. Her biggest worry is that her children might get hit by one of the Ferraris or Lamborghinis that her wilder neighbors like to zoom up and down the frond's single road.

It's easy to arrange such a Xanadu when you build the entire place, including the land it sits on, from scratch. On the Palm, everything is artificial except the air. The land you walk on, the desalinated seawater you drink, the imported foods you eat— everything was brought there and manufactured by human hands. And so much of it is sand: sand forms the ground under your feet, sand makes the walls around you, sand is in the plate glass sliding patio door that looks out onto your sandy beach.

"I just love it," said Hart. "It doesn't get any better than this."

In Dubai, the conversion of underwater sand into artificial land is making developers rich and wealthy buyers happy. But a few thousand miles away, the same process has spawned an extraordinarily dangerous confrontation between the world's two mightiest nations.

Five hundred miles off the southern coast of China is a hotly disputed patch of the South China Sea. Some 10 percent of the world's fish come from here, and perhaps more critically, billions of barrels of oil and trillions of cubic feet of natural gas lie under the seafloor.[29] It's also one of the world's busiest shipping routes. So it's no surprise that virtually every country in the region— China, Taiwan, Vietnam, Brunei, Malaysia, and the Philippines— lays claim to a scattering of rocks and reefs called the Spratly Islands that sit strategically in that area.

Since the 1970s, in an effort to bolster their claims, most of those countries have enlisted dredged-up sand from the seafloor to build up one or another of these tiny islands to a size that could accommodate an airstrip. These were relatively small additions; the biggest before 2014 was a reclamation project by Vietnam that added sixty acres of land to its outposts.[30]

Then China, which holds seven Spratly outcroppings (one of which it seized from Vietnam in a 1988 clash that left dozens of soldiers dead), decided to assert itself.

In recent years, along with its vast, state-directed expansions of road and rail networks, urban infrastructure, and practically every other aspect of its economy, China has also built up an armada of oceangoing dredging ships, among the biggest and most technologically advanced in the world. It buys some ships from abroad, but increasingly manufactures its own. The country's annual dredging capacity—the volume of sand and muck it can haul up from underwater—has more than tripled since 2000, to more than a billion cubic meters. That's more than any other nation.[31]

The pride of this fleet are enormous, technologically advanced craft called self-propelled cutter-suction dredges. A boom arm, capped with a cutter head—a large steel ball studded with teeth— protrudes from the bottom of these ships down into the seabed. The ball spins around, its teeth tearing up sand, rocks, and whatever else is down there, while a built-in pump sucks the grains up into the ship. The slurry is then shot through a pipeline floating on the water's surface, which can extend for miles, onto a reef or rock, where it piles up to create new dry land. China's mightiest dredge, the biggest in Asia, was launched in 2017. Dubbed "the magical island-maker," it can haul up nearly 8,000 cubic yards of sand and other material per hour from depths up to 100 feet.[32]

China has made the manipulation and movement of sand into a potent tool of statecraft. In late 2013, Beijing set a fleet of ships to work expanding its pieces of territory in the Spratly Islands. In satellite photos, the ships look like a flock of confused sperm[33] swimming away from the ova of the growing islands, the tails of their pipelines flailing behind them. Within eighteen months, these ships built nearly 3,000 acres of new land. That's seventeen times more land than all the other regional claimants have added to the islands in the past forty years combined.[34]

This de facto territorial expansion set off alarms from Manila and Hanoi to Washington, DC, for several reasons. For one, all that land-building has been calamitous for the nearby environment. Most of the coral and other life-forms on the seven reefs themselves were, of course, destroyed by the mountains of sand dumped on top of them. The dredging also churned up sand and other debris that clouded the waters for miles around, harming other nearby reefs that provided habitat for countless fish, as well as endangered giant clams, dugongs, and several species of turtle. In 2016, an international tribunal convened to address the Philippines' complaints about China's activities in the South China Sea produced a study that concluded "China's artificial island-building activities . . . have caused devastating and long-lasting damage to the marine environment."[35] An American marine biologist called it "the most rapid rate of permanent loss of coral reef area in human history."[36]

But even more disturbing than the new islands' environmental impact are their geopolitical implications. Almost as soon as the sand was dry, China began building military bases on the Spratlys.

The armed forces have installed antimissile weaponry, runways capable of handling military aircraft, structures that US officials believe are designed to house long-range surface-to-air missile launchers, and port facilities that may be capable of accommodating nuclear submarines. "This is extremely worrying for nearby countries," says Gregory Poling, an expert on the South China Sea with the Center for Strategic and International Studies. "They now have Chinese air and naval bases right next door. China is establishing de facto control, so that it won't matter what the international community says."

Beijing didn't stop with the Spratlys. It also built new territory in another tiny collection of South China Sea islands called the Paracels, where it installed airstrips and missile batteries and reportedly plans to deploy a floating nuclear power plant to provide power.[37] Meanwhile, in another sign of Beijing's ambitions to expand its global reach, in 2017 China opened its first overseas military base, in the African nation of Djibouti. That base didn't require any land reclamation, but future ones might. China's new power to alter geography with sand means that if necessary, it can change the shape of other friendly countries' coasts or islands to accommodate its warships.

The Spratly Islands have become a major flashpoint between China, the United States, and its Pacific allies. "China is building a great wall of sand with dredges and bulldozers," warned Admiral Harry Harris, commander of the US Pacific Fleet, in a 2015 speech.[38] China refused a US request to halt construction that year, declaring "the South China Sea islands are China's territory."[39] The Obama administration responded with air and naval patrols through the area.

The early days of the Trump administration, however, ratcheted up tensions to unprecedented heights. At his confirmation hearings,

Secretary of State Rex Tillerson compared China's Spratly buildup to Russia's invasion of Crimea.[40] He added: "We're going to have to send China a clear signal that, first, the island-building stops and, second, your access to those islands also is not going to be allowed." In response, state-run Chinese media warned that if the Trump administration was to try to blockade the islands, "it would set a course for devastating confrontation between China and the US."

Stephen Bannon, at the time one of Trump's key advisors and a member of the National Security Council, seemed to welcome that prospect. A few months before he officially joined Trump's campaign, Bannon told listeners to a radio show he hosted that China is "taking their sandbars and making basically stationary aircraft carriers and putting missiles on those." His conclusion: "We're going to war in the South China Sea in five to ten years. There's no doubt about that."[41] Bannon looked prescient in September 2018 when a Chinese destroyer trying to scare an American warship away from the Spratlys nearly collided with it.

The armies of sand may be pushing the human armies of the world's two mightiest nations closer to conflict. For all the ways it helps us, sand can also endanger us. Which brings us back to deserts.

INTERLUDE
The Fighting Arenophile

If we don't count Indiana Jones, there aren't many scientists who were also daredevil Nazi fighters—even fewer when it comes to scientists specializing in sand. In fact, there's probably just one. That would be scientist-soldier Ralph Bagnold, explorer of the Sahara, scholar of the physics of sand, and scourge of the Third Reich.

As a young British army officer posted to Egypt in the 1920s, Bagnold became fascinated with the desert. In his spare time, in fine mad-dogs-and-Englishmen style, he customized Model T Fords with oversize radiators, low pressure tires, and other modifications to enable them to drive in the sand, allowing him to explore deeper into the Sahara than any European had ever gone. He came to know the trackless terrain intimately.

Then World War II broke out. Britain's forces in Egypt found themselves facing off across the Sahara against Italian and German troops in Libya. Suddenly Bagnold's eccentric hobby became a potent weapon. With his unmatched knowledge of the desert, now-Major Bagnold was charged with creating an elite commando force. In September 1940, Bagnold's Long Range Desert Patrol Group, made up of a few hundred volunteers from England, New Zealand, Rhodesia, India, and other corners of the British Empire, went into action. "I had been given complete carte blanche . . . to make trouble anywhere in Libya," he later wrote.[42]

Bagnold's men traveled deep into the uncharted sands in trucks equipped with enough food, water, and ammunition to keep them going for weeks. They cultivated a desert-pirate look, sporting Arab headdresses, unkempt beards, and a scarab insignia. Camouflaged amid the dunes, miles behind enemy lines, they monitored troop movements, radioing their intelligence back to British forces in Cairo. They launched lightning surprise raids on Axis convoys and airfields, then disappeared back into the vastness of the Sahara. They guided Allied troops through what was thought to be impassable desert, enabling them to launch a surprise attack that played a key role in defeating the Nazi "Desert Fox," Field Marshal Erwin Rommel.[43]

After the Axis's African surrender in 1943, the LRDP went on to missions in Greece, Italy, and the Balkans before finally disbanding at the war's end. Bagnold's obsession with sand, however, continued. He became one of the world's foremost scholars on how sand moves, and wrote the definitive text on the physics of wind-blown sand and desert dunes. Bagnold died in 1990, but his research is still in use: NASA scientists consulted it in planning its missions to Mars. Not even Indy could claim that.

CHAPTER 9

Desert War

From the top of a certain windblown hill in Duolun County, in China's Inner Mongolia region, the view could be described as either profoundly inspiring or deeply strange. For miles around, the terrain is dun-colored and dry, sandy desert stubbled with yellow grass. But the cluster of hillsides closest to the one I found myself standing on in spring of 2016 were emblazoned with enormous, carefully configured swatches of green trees. They were planted to form geometric shapes: a square, a hollow-centered circle, a set of overlapping triangles. The flatland below them was striped with ruler-straight bands of young pine trees, all the same height, standing in formation like soldiers ready for battle.

Zuo Hongfei, the cheery deputy director of the local "greening office" of China's State Forestry Administration, eagerly pointed to an eighty-foot-long display showing how barren this part of Duolun County was just fifteen years ago, before a massive greening campaign installed millions of trees across the land. Photos and satellite images show it was largely desert, dotted here and there with spindly trees and shrubs. "See?" said Zuo, pointing out a picture of an old man and a young girl in front of a low dwelling half swamped by dunes. "The houses were almost buried by sand!"

Though armies of sand are our indispensable allies, supporting our way of life in so many ways and in so many places, they can also turn against us, becoming a remorseless enemy force. The vast legions of sand in the world's deserts are largely useless when it comes to building cities; in some places, as if angry at being left out, they have become threats to those cities.

The sand lands that cover about 18 percent of China have expanded rapidly. By 2006, they were devouring usable land at a rate of almost 1,000 square miles per year, nearly the area of Yosemite National Park, up from 600 square miles[1] per year in the 1950s.

That's a problem not only for the people living in those areas, but also for the many millions more who live close enough to deserts to be affected by the movements of sand. Migrating dunes threaten farm fields and even whole villages. Stretches of roads and railways are constantly shut down by blown sand. Sandstorms regularly blow tens of thousands of tons of sand and dust into Beijing and other cities, snarling traffic and creating a vicious health hazard. The World Bank has estimated that desertification costs the Chinese economy some $31 billion per year.[2]

This is an issue that goes far beyond China. According to the United Nations, desertification directly affects 250 million people worldwide, including parts of the United States.[3] Sand is slowly burying the once-flourishing Malian town of Araouane, on the edge of the Sahara Desert. In 2015, a massive sandstorm blanketed Lebanon and Syria, killing twelve people and sending hundreds to the hospital with respiratory problems. And particles from dust storms in China have clouded the air as far away as Colorado.

Deserts have always advanced and retreated over the centuries, driven by large-scale shifts in atmospheric and geologic conditions. But what's happening in our time is different. It's not that the world's deserts are spreading like some aggressive disease; rather, the land surrounding them is drying out.

Climate change, by raising temperatures and reducing soil moisture, is partly to blame. But the main culprits are people. Lots of people. The population of Inner Mongolia, where much of China's desert lies, has quadrupled in the last fifty years to more than 20 million, mostly thanks to ethnic Han Chinese moving into the area. Those people cut trees for firewood and draw groundwater to irrigate farmland and run heavy industries. The number of livestock has also grown sixfold, and those animals eat a lot of grass. As underground aquifers get depleted, the land dries up. Without plant roots to anchor it and moisture to weight it, topsoil blows away, leaving behind only pebbles and sand. Which means that at the same time that we're running out of the sand we need, we're generating more of the kind we don't.

"We can probably go on for another five years, possibly ten, but after that it's simply not an option to go on losing land at the present rate," Louise Baker, a senior adviser to the UN Convention to Combat Desertification, told a British newspaper. "Every minute, twenty-three hectares of land are lost to drought and desertification. The global population is already 7 billion, and by 2050 it's projected to reach 9 billion. We need to produce more food, but the area of productive land is going down every year."[4]

Duolun County, which lies at the southern edge of the Gobi Desert, has always been a dry place. But during the last century, decades of overfarming and overgrazing desiccated huge areas of it into pure desert. By 2000, 87 percent of its total area was sand land. The situation was so dire that in 2000 Premier Zhu Rongji visited the area and declared, "We must build green barriers to block sand."

And so they did. In the first fifteen years of this century, the government planted millions of pine trees all over Duolun County. More are put in the ground every spring. Zhu's "green barriers" aren't just blocking the sand; they're forcing it to retreat. By now, according to official Chinese statistics, 31 percent of Duolun's land is forested. The total would be even higher but for some missteps in the early years of the project, says Zuo, the county greening officer. Huge numbers of fast-growing poplars were planted, but most of them died. "We had poor knowledge then," says Zuo. "We found they needed too much water."

Duolun's afforestation project is just a tiny sliver of a project of bedazzling scale unfolding across the country. China is building a new Great Wall—this one aimed not at repelling invading Mongols, but a more insidious menace from the northern drylands. This wall is being built not of stone but of trees—*billions* of trees, enough to stretch nearly the distance from San Francisco to Boston. Its purpose: to push back China's vast deserts.

The project, officially dubbed the Green Great Wall, was launched in 1978, and is slated to continue until 2050. It aims to plant some 88 million acres of protective forests, in a belt nearly 3,000 miles long and as wide as 900 miles in places. Prompted by China's ever-worsening environmental conditions, the government has added a handful of other major afforestation projects in more recent years. It all adds up to what is easily the biggest tree-planting project in human history.

The results so far have been splendid—at least according to the Chinese government. Thousands of acres of moving dunes that threatened farmers' fields and villages have been stabilized. The frequency of sandstorms nationwide fell by one-fifth between

2009 and 2014. And though deserts continue to spread in some areas, the State Forestry Administration, the government agency that oversees the main tree-planting programs, claims that on balance it has not only stopped but even begun to reverse the deserts' expansion.[5]

It's heartening to see a nation famous for its warp-speed industrialization and world-beating levels of pollution undertaking such a colossal effort to make their nation green. But many scientists in China and abroad say the actual results are unimpressive at best and disastrous at worst. Many of the trees, planted in areas where they don't grow naturally, simply die after a few years. Those that survive can soak up so much precious groundwater that native grasses and shrubs die of thirst, causing more soil degradation. Meanwhile, the government has forced thousands of farmers and herdsmen to leave their lands to make way for the desert-fighting projects.

In short, China has undertaken the most ambitious effort anywhere to beat back the sands of the desert, and it appears to be winning. But that victory raises some troubling questions. What is the cost it has incurred—and will it last?

China isn't the first country to try shoring up degraded lands with man-made forests. In the 1930s, the US government under President Franklin D. Roosevelt planted some 220 million trees in a largely successful effort to block the dust storms blighting many central American states. Joseph Stalin launched a similar effort in the 1940s, planting more than 10,000 square miles of steppe land with trees; almost all of them were dead within twenty years.[6] Algeria tried planting a 930-mile "green dam" in its southern desert in the 1970s, with lackluster results.[7] And today in Africa, eleven countries are fitfully trying to create a continent-wide green barrier similar to China's to hold back the spreading Sahara. As in

China, the problem is largely driven by demography: the population of the Sahel, the semiarid region bordering the Sahara, has more than quintupled in the last sixty years.

But nothing touches the scale of China's sylvan crusade. Practically since the Communist Party took power in 1949, it has promoted tree planting as a righteous cause, even a civic duty. Tree planting kicked into overdrive with the launch of the Green Great Wall in 1978, the same year Beijing began opening up the Chinese economy. Since the project's inception, Chinese citizens have planted billions of trees, foresting an area larger than California.

One major reason China has been able to get so many trees in the ground so fast is the same reason it has been able to open so many factories so fast: by freeing people to make money. Rather than relying on revolutionary idealism, the government now pays villagers to plant trees. In some places, the government also leases their land for afforestation. Entrepreneurs cultivate and sell seedlings to the government, and harvest mature trees for lumber. According to official Chinese statistics, all of this has reduced poverty in many areas. It has also made a few people very rich.

Wang Wenbiao is one of those people. He grew up in a village on the edge of Inner Mongolia's vast Kubuqi Desert, adjacent to but not technically part of the Gobi Desert, in a family of farmers so poor he and his siblings were allotted one new set of clothes per year. They were on the front lines facing the adversary of sand. Wind constantly blew grains into their bed and onto their food. "Two words were very important in my childhood," says Wenbiao. "Sand and poverty."

Sand is still an important part of Wang's life, but the poverty is long gone. These days, he runs a multibillion-dollar[8] corporation that aims to not only help hold back the desert but also make a profit from it.

I met Wang one spring morning in the sleek Beijing headquarters of Elion Resources Group, the putatively environmentally beneficent enterprise he heads. The vibe was imperial. Wang is a mirthless, heavyset, middle-aged man, his thick hair swept back off his broad forehead. He was seated in a white leather chair in front of a mural depicting waterfalls and forests. Arrayed around him on more white chairs were me, my interpreter, a company PR rep taking notes on everything, and another aide who chimed in frequently to reinterpret how my interpreter had interpreted Mr. Wang's declarations.

Wang got his start at age twenty-nine when he was appointed head of a salt and mineral mine in the Kubuqi Desert, in northeastern China. Sand bedeviled him from his first day on the job. "A jeep took me to the mine, but it got stuck in the sand outside the gate," he recalled. "Rather than give me a proper welcome, the workers had to come and help me get out." Sand and transport, Wang realized, were his biggest problems. There was no direct road from the factory to the outside world. The salt field sat only 37 miles from a railway station, but reaching it required a 200-mile detour. With funding from the local government, Wang set to work building new roads and planting trees and shrubs alongside them to keep the sand from inundating them. By now the company has planted 30 percent of the Kubuqi Desert—some 2,300 square miles—a feat that has earned it recognition from the United Nations.

The barriers kept the roads passable, and the salt factory's business boomed. Wang's company branched out into other industries, including chemicals and coal power plants. Today it employs over 7,000 people. It is now seeking to rebrand itself as an eco-friendly enterprise, singing a song sure to please the ears of modern investors concerned about the environment. The company runs solar power fields, cultivates licorice and other desert plants prized in traditional Chinese medicine, and claims to bring thousands of ecotourists to the Kubuqi every year. Elion has also become a major contractor for the Green Great Wall, installing instant forests from the western deserts to an area north of Beijing that will host the 2022 Winter Olympics.

"Green land and green energy," says Wang. "That will be our future direction." When pressed, though, he acknowledges that about half of the company's $6 billion in annual revenues still come from "traditional" industries, including chemical production and coal power plants.

Elion's flagship project is its tree-planting campaign in the Kubuqi Desert. The word *desert* is often used loosely, a judgmental label slapped on a whole range of low-moisture drylands. The Kubuqi isn't your American Southwest, Palm Springs–type desert, drylands bedizened with cactus, creosote, and Joshua trees. The Kubuqi is mostly sand, and nothing but sand.

A trip along one of the Elion-built roads through it was surreal, almost dreamlike. The road was a ribbon of smooth asphalt lined on both sides with orderly ranks of stubby pine trees and slender poplars, spears of green sticking straight up out of the sand. Elion billboards in Chinese and English popped up every couple of miles trumpeting eco-corporate-Communist slogans: *Promoting Eco-Civilization; Green Desert—Beautiful China; Ecology brings benefits, green brings prosperity.* Most of the trees were no taller than a fifth grader; the bulk of them have been planted only during the last few years. Past those belts of green, as far as the eye could see there was nothing but barren, rolling sand dunes.

The road eventually led to the company's palatial, dome-topped "Seven Star Kubuqi Hotel." It was surrounded by carefully irrigated rows of poplars and swaths of green grass, with a fountain out front. The hotel grounds include, improbably, a golf course. When a hotel staff member spotted my photographer Ian Teh out there one day, he hurried out and demanded Teh delete his pictures.

How can a desert sustain so many trees, let alone a golf course? Where does all the water come from? "Everyone asks this question," replied Wang with a gruff fraction of a smile. The trees use only a tiny amount of the region's underground water, he claimed; the most important factor is that the company has literally made it rain. Increased evaporation from all the new plants has made the climate more humid, Wang declared. "Twenty-eight years ago, there was only about 70 millimeters of rainfall. In recent years it has reached 400 millimeters," said Wang. "We changed the ecosystem."

I asked several independent Chinese and international researchers about this claim. All of them were skeptical. Planting an area that large might increase humidity and rainfall to some extent, they agreed, but to more than quadruple it? "Sounds like bullshit to me," said Mickey Glantz, a University of Colorado researcher who has been studying deserts in China and around the world for forty years.

Cao Shixiong, a lean, banty researcher at Beijing Forestry University, has a simple explanation. "When there's profit at stake, people tell lies. The central government gives out billions of yuan every year for tree planting. So there are many companies that want to take part. They're not concerned with the environment, but with profit."

Cao used to be a believer. He spent twenty years working on State Forestry Administration tree-planting projects in Shaanxi Province. "I thought it was a very good way to combat desertification," he said. But his trees never survived for long. "I realized it's because of policy. The problem is, we were choosing the wrong place to plant trees."

Cao and most critics of the tree-planting campaign acknowledge that it has benefited some areas. But those benefits, they argue, are localized, and may not last. In some ways, they may even be making things worse.

It's true, for instance, that sandstorms have decreased around Beijing in recent years, a welcome development for which some researchers credit the Great Green Wall. Other experts, however, say that change may be at least partly because there's been more rain in northwestern China over the last several years, which keeps dust down and makes more plants grow naturally.

"Nobody knows how much is because of the government and how much is natural," said Shen Xiaohui, a retired SFA engineer. "But you know the government will claim it's all because of them."

It's also undeniable that billions of trees have been planted in formerly barren areas, and in some places those artificial forests are thriving—stabilizing and enriching the soil and generally making those areas more livable. But huge numbers of them have also died.[9] Some fell prey to the arid environment, some to diseases and pests that spread rapidly through the monocultural artificial forests. In 2000, a beetle infestation in north-central China wiped out 1 billion poplars, the fruits of two decades of planting.[10]

The most serious concern is that all those newly planted trees will suck up the desert's precious groundwater. That's what's keeping millions of them alive at this point. In Duolun County, Zuo Hongfei assured me, this isn't a problem, because the area naturally gets enough rain to sustain the drylands-adapted trees they've been careful to plant.

But research suggests it is already happening in other, drier parts of China. Ultimately, that could cause not only the trees but also whatever smaller plants grew there naturally to die of thirst, leaving the land in worse shape than ever.[11] "For the past thousand years, only shrubs and grass have grown in those areas. Why would you think planting trees would be successful?" asked Sun Qinqwei, a former researcher at the Chinese Academy of Sciences' Desert Research Institute who now works for the Washington, DC-based National Geographic Society's China Program. "You can succeed in the short term by pumping groundwater, but it's not sustainable. Investing money on trees that are not supposed to be there is kind of crazy."

So what's the bottom line? Is the Green Great Wall hurting or helping? It's hard to know. The effects of such a large and complex change to the environment can take years, even decades to manifest themselves. In the meantime, considering the enormous scope of these programs, good data is startlingly scarce. As a 2014 study[12] of China's major tree-planting programs by a group of American and Chinese scientists concluded, "the extent to which the programs have changed local ecological and socioeconomic conditions are still poorly understood, as local statistics . . . are often not available or unreliable." Another study by the Chinese Academy of Sciences and Beijing Normal University adds: "Although numerous Chinese researchers and government officials have claimed that the afforestation has successfully combated desertification and controlled dust storms, there is surprisingly little unassailable evidence to support their claims."[13]

Factor in also that, for Chinese researchers at least, criticizing a pet project of the autocratic government carries real risks. Cao says that's the reason he hasn't been able to get any outside research money for the last five years. "Before my academic career, I thought science was just science," he said. "But science is nothing when facing politics."

On the other hand, bureaucrats and researchers connected to the State Forestry Administration all have plenty of reason to declare the Great Green Wall a roaring success. "There are stakeholders all along the chain," said Sun. "There are State Forestry Administration bureaucrats in every province and county. They get a lot of money for planting trees." Considering that the SFA is tasked with both planting millions of trees and assessing whether it's a good idea to plant millions of trees, you can understand why outsiders are skeptical of their findings.

A few miles from the Duolun County hilltop with the view of all those new trees lies a settlement called New Granary Village. It's a grim assemblage of small, cookie-cutter brick houses lining a grid of bare dirt streets, most of them unrelieved by so much as a blade of grass. It calls to mind less a village than a long-term refugee camp—which, in a sense, it is. New Village was built early in this century to house some of the 10,000-plus local farmers who have been forced to relocate to get them out of the way of the SFA's new trees. They are some of the hundreds of thousands of mostly Mongol, Kazakh, and Tibetan farmers and herders whom the Chinese government has forced to move off the grasslands and into urban areas, leaving their traditional way of life behind. Officially this is to reduce overgrazing. Many believe it's also a land grab to free up water and other resources for Han Chinese businesses. In some places the herders have resisted with violent protests.

"We didn't want to move, but we were forced to. They would have demolished our home if we had stayed," said Wang Yue, a sinewy sixty-five-year-old with a resigned air. He was born and raised just a few miles away, in a now-vanished village where his family had lived for generations. He has a decent house in New Granary Village—a couple of rooms with a platform to sleep on, a coal stove to cook on, and windows looking out on a tiny courtyard. But he lost his land when he moved. "Life was better in the old village," he said. "Here we have to buy oats to feed the animals. We used to just let them graze." He ekes out a living now doing odd jobs, but at his age, it's getting difficult. His wife is dead, and his two daughters have moved away. He said he has never received the government subsidies he was promised, a complaint I heard from several others in New Granary Village.

"They lied to us," he said. "Tree planting is making some officials rich, but we lost so many things."

Desert sands and their own government combined to force Wang and his neighbors from the rural villages of their ancestors into an urban-style settlement. That's a story specific to Inner Mongolia. But the experience of migrating from a rural village to something resembling a city is one he shares with hundreds of millions of people. That migration is rapidly reshaping our world, in ways that are forcing humans to rely ever more heavily on ever larger armies of sand.

CHAPTER 10

Concrete Conquers the World

A thousand miles southeast of Duluon County stands the gleaming megalopolis of Shanghai, China's biggest city and main financial center. Thirty years ago, almost everyone in Shanghai lived in two- or three-story *shikumen*, picturesque alleyway complexes fronted by stone gateways.[1] But the shikumen have all but disappeared, bulldozed out of existence in the ongoing maelstrom of development that has transformed the city since the 1990s.

Shanghai's growth makes Dubai's look halfhearted. Seven million new residents have poured into the city just since 2000, raising its population to more than 23 million.[2] In that time, Shanghai has built more high-rises than there are in all of New York City. That's on top of adding countless miles of roads, a gargantuan international airport, and other infrastructure.[3]

Manufacturing all the concrete needed to create this urban colossus has required mobilizing armies of construction sand on an unprecedented scale. In the early years of Shanghai's boom, much of the sand for all those new buildings and roads was recruited from the bed of the Yangtze. Miners—many of them operating illegally—pulled out so much sand that bridges were undermined, shipping was snarled, and 1,000-foot-long chunks of riverbank collapsed.[4] Unnerved by the damage being done to the nation's most important river, which provides water to some 400 million people, Chinese authorities banned sand mining on the Yangtze in 2000. That sent the miners flocking to Poyang Lake, China's largest body of freshwater, which drains into the Yangtze some 400 miles from Shanghai.

Now, on any given day, hundreds of dredges, some the size of tipped-over apartment buildings, may be on the lake. The biggest can haul in as much as 10,000 tons of sand per *hour*. A study by a group of American, Dutch, and Chinese researchers estimates that 236 million cubic meters of sand are taken out of the lake annually. That makes Poyang Lake the biggest sand mine on the planet, far bigger than the three largest sand mines in the United States combined.

Pulling those legions of grains from the bottom of Lake Poyang is a profitable business, but it may be wreaking havoc on the lake itself. Poyang's water level has been dropping dramatically in recent years, and researchers believe sand mining is a key reason. The dredging boats haul out thirty times more sediment each year than the amount that flows in from tributary rivers, according to David Shankman, a University of Alabama geographer and one of the authors of the study that came up with the figure of 236 million cubic meters. "I couldn't believe it when we did the calculations," he said. So much sand has been scooped out, Shankman and his colleagues believe, that the lake's outflow channel has been dramatically deepened and widened, nearly doubling the amount of water that flows out into the Yangtze.[5]

The resulting lower lake levels translate into changes in water quality and supply to surrounding wetlands that could be ruinous for the lake's inhabitants. Poyang Lake is Asia's largest winter destination for migrant birds, hosting millions of cranes, geese, storks, and others—including several endangered and rare species—during the cold months. It is also one of the few remaining habitats for the endangered freshwater porpoise. Researchers warn that besides the loss of habitat, the sediment stirred up and noise generated by sand boats foul up the porpoises' vision and sonar so badly they can't find the fish and shrimp they feed on.

To make matters worse, as several local fishermen told me, there are fewer fish to be found in the first place. "The boats are destroying our fishing areas," said one fifty-eight-year-old woman, who did not want her name published. The dredging, she explained, destroys fish breeding grounds, muddies the water, and tears up her nets. She lives in a village on the lakeshore that is little more than a tiny collection of ramshackle houses and battered wooden docks. It is dwarfed by an offshore flotilla of dredges and barges with industrial cranes jutting from their decks.

In the twenty-first century, the army of sand has fanned out to conquer the entire world. The building methods and materials that a hundred years ago were mostly confined to wealthy Western nations have in the past thirty spread to virtually every country. This epochal shift is what lies behind the sand crisis.

Though we use sand for thousands of purposes, concrete is really driving that crisis. More sand particles are pressed into service to make concrete than all those used for asphalt, glass, fracking, and beach nourishment put together. Poyang Lake, Morocco's beaches, Kenya's rivers, the fields outside Paleram Chauhan's village—they're all being pillaged to make concrete.

Concrete has become the most widely used building material

on Earth; we use twice as much concrete every year as steel, aluminum, plastic, and wood combined. An estimated 70 percent of the
world's population live in structures made at least partly out of
concrete.[6] The world's biggest dams and bridges are all made with
reinforced concrete. Even steel-framed skyscrapers require massive
quantities of concrete for their foundations and floors. The earth's
total paved area is estimated to be more than 223,000 square
miles—just a bit less than the entire state of Texas.

"The equivalent of forty tons of this material exists for every
person on the planet," writes historian Robert Courland in *Concrete Planet*. "And an additional one ton per person is added with
every passing year."[7]

The reason we are using so much concrete, as we've seen, is the
historic demographic shift that is changing how people live in almost every country on Earth: urbanization. Every year, tens of
millions of people, especially in the developing world, leave the
hardship and poverty of rural villages for a shot at a better life in
the city.

Across Africa, the Middle East, Latin America, and especially
Asia, towns are swelling into cities, and cities are bloating into
megacities. In 1990, there were only ten cities in the world with 10
million inhabitants or more. By 2014, there were twenty-eight of
them, home to a total 453 million people.[8] Those people want the
armies of sand to work for them, too. They want and are getting,
however unevenly, the benefits of concrete and glass homes, offices, shops, and roads. Even places that used to be completely
empty of people are now thick with concrete high-rises and paved
roads, from Dubai to Inner Mongolia.

We are building cities so fast that "the volume of urban construction for housing, office space, and transport services over
the next 40 years could roughly equal the entire volume of such

construction to date in world history,"[9] according to the US National Intelligence Council.

There's no way cities could grow this fast without sand, in the form of concrete. It's an almost supernaturally cheap, easy way to quickly create relatively sturdy, sanitary housing for huge numbers of people. It's strong, capable of holding thousands of tons worth of people, furniture, and water. It won't burn or get infested with termites. And it's incredibly easy to use. A single person can mix a batch of basic concrete and slap together a serviceable shelter. A well-financed contractor can pour the foundation of a towering building in a matter of days.

Urban areas are mushrooming everywhere, but China is on a city-building spree that beggars anything the world has ever seen. There are more than 220 Chinese cities with over a million inhabitants; the entire continent of Europe has only 35. Upwards of half a billion Chinese now live in urban areas, triple the total of sixty years ago.[10] That's about the same as the total combined populations of the United States, Canada, and Mexico. And millions more come every year.

To connect all those urban centers, China is also vastly expanding its road network, as well as its airports and maritime ports. To help power them, it is building dams, including the infamous Three Gorges Dam, the biggest civil engineering project in history, a leviathan comprising more than 35 million cubic yards of concrete.[11] Meanwhile, Chinese companies are building thousands of miles of roads and hundreds of high-rises all over the world, from central Africa to central Europe.

China is so building-happy that in recent years whole cities have been built from scratch that aren't even needed—at least not yet. Filled with uninhabited apartment blocks and unused offices, they've become known as "ghost cities." Most are in China's relatively poor and undeveloped western regions. The government invests in them in hopes they will lure people away from the country's overcrowded eastern shore, and developers build them thinking they will be cash cows. Actual residents, however, are proving slow to take the bait.

The "city" of Kangbashi, for instance, sits on the edge of the Inner Mongolian desert. It was built from scratch in 2004. Architecturally speaking, it's impressive, or at least ambitious. It features a meticulously landscaped central plaza more than a mile in length, along which sits a library shaped like a trio of enormous shelved books, a museum shaped like a cross between a peanut and a bronze beanbag, and an art gallery vaguely modeled on a pair of yurts. Wide avenues lead to shopping malls, hotels, and high-rise housing developments. The city was built to house more than a million residents.

But when I was there in spring of 2016, it held barely one-tenth that number. On a Thursday afternoon, the only people in the plaza besides my interpreter and me were a scattering of cleaning workers lazily ambling after the odd bit of windblown trash, and a solitary pedestrian in the distance. The shopping-mall-sized library was dark and nearly deserted. As I was on the way into the library's main entrance, my phone and camera set the walk-through metal detector buzzing angrily. No one stirred.

All of this frenetic construction has made China into the world's largest consumer of concrete[12] and the most voracious consumer of sand in human history. In 2016, China used an estimated 7.8 billion tons of construction sand. That's enough to cover the entire state of New York an inch deep. In the next few years, that number is projected to grow to nearly 10 billion tons.

All around the world, converting those armies of sand into concrete has in many ways yielded incredible blessings. Concrete has saved countless lives and enriched even more. Concrete dams generate electricity. Concrete hospitals and schools can be built and repaired far more quickly than their counterparts of adobe, wood, or steel. Concrete roads help farmers get their crops to market, students to get to school, sick people to get to hospitals, and medicines to get to villages in all weathers. Research has shown that paving streets increases land values, agricultural wages, and school enrollment.

Just having a concrete *floor* is a huge improvement for many people. Hundreds of millions of people worldwide live in dwellings with dirt floors. Writing in *Foreign Policy* magazine,[13] economist Charles Kenny pointed out that walking barefoot on such a floor is an excellent way to contract an illness, particularly hookworm disease, a parasitic infection to which children are especially vulnerable. Simply paving those floors massively reduces the risk. According to Kenny, a program in Mexico that provided concrete floors for poor homes cut the rate of parasitic infestations by nearly 80 percent, and halved the number of children with diarrhea in any given month. Sand, it turns out, can not only provide shelter but can be a boon for public health.

All of these countless tons of concrete, however, come at a steep cost. Several types of costs, actually.

Heaping concrete on cities can destroy culture and beauty just as surely as heaping sand on coral reefs kills fish. Shanghai's shikumen are hardly the only historic architectural type demolished to make room for concrete high-rises. Concrete is a key reason so many places in today's world look just like every other place. It's the standardized substrate upon which a million identical office towers, apartment blocks, Starbucks, Marriotts, and eight-lane highways have been propagated the world over. It is the coat of generic gray paint that renders everything the same color and texture. Sure, concrete has some cachet in certain architectural circles, but for the average person it's the symbol of modernity at its worst, the stuff they used to pave paradise and put up a parking lot.

More pressingly, concrete also inflicts physical harm on people and the planet. Just as it does on a beach, sand in the forms of concrete and asphalt soaks up the sun's heat. Those miles of warmed-up pavement can raise the ambient temperature of a whole city, creating a phenomenon known as urban heat islands. According to a 2015 study by the California Environmental Protection Agency,[14] when combined with the heat generated by motor vehicle engines, paved areas can boost the temperature in some cities by as much as 19 degrees Fahrenheit. That's more than just unpleasant. Heat exposure can be lethal to children, the elderly, and other vulnerable people. Heat also boosts the formation of air pollutants, especially ground-level ozone, better known as smog. Too much sand on the ground can lead to toxins in the air.

Urban heat islands will only get hotter as climate change grows more acute. And speaking of climate change, concrete is making it worse. The cement industry is one of the world's leading sources of greenhouse gases. Processing limestone into cement emits carbon dioxide. On top of that, most cement-producing furnaces burn fossil fuels, which spew out even more CO_2. Cement is made in at least 150 countries, and produces between 5 and 10 percent of the total carbon dioxide emissions worldwide. That puts cement making in the top three sources of carbon dioxide emissions, behind only coal-fueled power plants[15] and the ubiquitous automobile.

Concrete, as we have seen, is also the handmaiden of the automobile. One promotes dependence on the other. The more roads you build, the more traffic you generate, which means more carbon emissions from tailpipes. "Not to mention," as Charles Kenny writes, "building new roads in a pristine forest is a pretty surefire way to lose that forest to loggers."

In some places, building with concrete is backfiring in startling ways. Some 30 percent of Texas's Harris County, in which Houston sits, is covered by roads, parking lots, and other structures; that made the flooding caused by 2017's Hurricane Harvey much worse.[16] All that impervious concrete blocked storm water from seeping into the earth as it would naturally, turning streets into artificial rivers.

While concrete seals off the earth in Houston, it is crushing it in Indonesia. That nation's capital, Jakarta, and its environs are an urban behemoth of 28 million people, many of them living in the forest of skyscrapers that have sprung up in recent years. But the ground the city sits on is porous and weakened by the extraction of too much water by thirsty residents. As a result, the unfathomable weight of all that concrete is slowly squashing the ground beneath it, making the city sink. Jakarta has sunk by thirteen feet over the past thirty years, and is still dropping three inches per year. Nearly half the city now sits below sea level, protected only by aging sea walls.[17] Shanghai and other cities are similarly crushing the ground beneath them.

Perhaps the most frightening aspect of our dependence on concrete is that the structures we build with it won't last. The vast majority of them will need to be replaced—and relatively soon.

We tend to think of concrete as being permanent, like the stone it mimics. In its early days, modern concrete was touted as a completely fireproof and earthquake-proof material, one that would never require repairs. "It has made possible a structure which is a guarantee of its own durability, as concrete improves with age," trilled *Scientific American* magazine in 1906.[18] That same year the *San Francisco Chronicle* marveled at a new concrete bridge over the San Joaquin River, declaring "the remotest generation of mankind will never have to construct another bridge at that place."[19] Ernest Ransome himself wrote that "the general wear and tear on a well-constructed reinforced concrete building is insignificant and confined to the finish coat of the floor."[20]

None of that turned out to be true. Concrete fails and fractures in dozens of ways. Heat, cold, chemicals, salt, and moisture all attack that seemingly solid artificial stone, working to weaken and shatter it from within.

"The disease that will kill your concrete depends on where you live," said Larry Sutter, a professor of materials science at Michigan Technological University. In his state, it's the winter cold. Concrete is microscopically porous, so a little water always seeps in. That water expands when it freezes, which can crack the concrete. The chemicals used to deice roads also degrade their concrete surfaces.

In Florida, the number one problem for concrete is corrosion of the internal rebar, caused by salt in the atmosphere. In California, it's attack by sulfates in the water, "which can turn concrete into mush in a couple of years," said Sutter. Other potential worries include bacterial and algal growth in humid areas, and acid deposition from pollution in cities. Underground concrete structures like water pipes, storage tanks, and even missile silos have to contend with damaging chemicals that filter down through the earth.[21]

One of the most pervasive threats to concrete is something called the alkali-silica reaction, which was discovered in 1940. It's caused by certain types of sand—silica—which when combined with alkali and water in the cement react to form a gel that can expand and crack the concrete from inside. It's a particularly pervasive problem, found on every continent except Antarctica. In 2009, cracks caused by ASR were found in the walls of a nuclear power plant in New Jersey. Concrete in at least two other nuclear plants has cracked seriously in recent years, according to the US Nuclear Regulatory Commission[22]; one was damaged so badly it ultimately had to be closed down.

By now builders have developed tactics to prevent ASR, most commonly by including fly ash in the concrete mix. "But there's lots of concrete already in place that's susceptible," said Sutter. And there's probably more being put in place. "There are areas in the United States where we have mined all the good aggregate, and so we're using stuff we would not have used twenty years ago," an aggregate industry consultant told me—that is, sand and gravel that is susceptible to the alkali-silica reaction.

Reinforced concrete is also made vulnerable by the very component that makes it so strong: the steel rods inside it. "Cracks that occur in a structure may be repaired, but not before air, moisture, and many other possible chemicals seep into the form to cause rust," Courland writes.[23] "As the rebar rusts, several things happen. Not only is the amount of 'good' steel reduced, but the diameter of the rebar expands to as much as fourfold its original diameter, causing more cracks and, in due course, pushing out chunks of concrete." Usually the slow-spreading damage is spotted and the building fixed or condemned, but in the worst cases, the structure may be so badly damaged it collapses.

When a dam or a twenty-story office tower or a parking garage starts to show that kind of stress, the owners call in an outfit like Chicago-based Wiss, Janney, Elstner Associates. WJE specializes in figuring out what's going wrong with concrete in everything from nuclear power plants to skyscrapers. Its engineers head to the trouble zone with ground-penetrating radar and other sophisticated imaging gear and bore out core samples of the concrete. These men and women have dangled from the tops of skyscrapers and rappelled down the Washington Monument and the St. Louis Arch in pursuit of samples containing information about those structures' health.

At WJE's sprawling headquarters north of Chicago, petrographer Laura Powers examines those concrete samples under a powerful microscope to determine, among other details, the quality of the sand used to make it. Powers is a serious arenophile, a sand lover; she collects samples of grains from all over the world, and likes nothing more than to talk about their different qualities. She is often called on to testify in court cases in which contractors are being sued for using substandard aggregate—sand or gravel that was the wrong size or shape, or contained reactive agents that can cause ASR. "We do a lot of evaluations of aging structures," said Powers. "What worries me are the structures we're *not* evaluating."

Concrete making has developed into a highly sophisticated science to meet the panoply of uses for which it is called upon. There are thousands of different types and mixes of concrete, each with specific properties tailored to specific purposes. The strength required for a chunk of suburban sidewalk, for instance, is very different from that required of a slab of dam holding back a river. Adding chemicals or fibers can make concrete lighter, faster curing, more flexible, resistant to corrosion, or pretty to look at. You might need to add retarders to slow down hardening in hot weather, or accelerators to speed it up in the cold, or superplasticizers to make it more fluid. You might add steel fibers to increase the concrete's impact resistance, or polypropylene fibers to help keep it from cracking.

Of the utmost importance is deploying the right sand and

gravel, the particles that make up the bulk of any concrete. Changing the size, shape, properties, and proportions of the aggregate in the mix gives you concretes of differing strength, durability, ease of use, and cost. Getting the right grains for the job is so important that in 2010 the US military was forced to import sand from Qatar to Iraq.[24] There's certainly no shortage of sand in Iraq, but the local granules weren't good enough to make the concrete needed to build protective blast walls around government ministries and other important structures.

WJE helps builders develop concrete mixes for specialized purposes. Its campus hosts a warren of labs where they put slabs, cylinders, and chunks of concrete made with various sands from around the world through stress tests simulating its real-world environment. The most punishing testing is carried out by John Pearson, the lean, brush-cut manager of WJE's cavernous structural testing lab. The lab contains a trailer-truck-sized steel frame, fitted with a hydraulic press capable of exerting 2 million pounds of pressure. WJE researchers use it for testing structural columns. Pearson showed me a video of a recent test. Second by second, as the press applied unimaginable force to a twenty-foot concrete column, fist-sized chunks of concrete started to pop loose. Then suddenly the column exploded in a burst of debris and dust, knocking the camera over. "That kind of sudden failure wouldn't happen in the real world, except maybe in an earthquake," Pearson explained. "But slow, gradual deterioration, if it's not noticed, or not addressed, can lead to collapse."

Edwin Mah spends his days looking for just such slow, gradual deterioration. Mah, sixty-seven, is a freeway bridge inspector with Caltrans, California's state transportation agency, charged with checking out how well bridges carrying millions of cars are holding up. He has a lean face with a toothy smile and an accent from his native China, which he left back in 1960. I joined him recently for an inspection of a typical bridge, one built in 1950 to carry the 101 Freeway over Melrose Avenue in central Los Angeles.

It's a grimy, dusty, noisy corner of the city. The summer heat was just kicking in at 8:30 A.M., and traffic streamed steadily on and off the ramps connecting the freeway with Melrose, a busy four-lane thoroughfare. Underneath the overpass, a bluntly functional span held up by two heavy concrete columns, were vestiges of homeless encampments: an abandoned shopping cart, scattered clothes, a mattress, ashes from a fire. Homeless folk add another risk to concrete overpasses, Mah said. They sometimes steal the steel nuts on the structure to sell as scrap metal, or accidentally set fire to wood reinforcements with their cooking fires. Caltrans inspectors always go out in pairs, Mah explained, especially to spots where homeless folk stay. "A lot of them are very rude," he says. Sometimes he has to call in California Highway Patrol officers to get them to move so he can do his job.

Mah climbed the slope from the street and stepped out onto the narrow shoulder of the bridge. A relentless fusillade of cars and trucks roared past no more than a foot from him, but Mah didn't seem to notice. As we walked, he pointed out cracks in the concrete road surface that had been filled in with tarry black sealant, and divots created by spalling—spots where internal expansion had popped off chunks of the concrete, exposing the rebar.

"You see this crack right here? This is very severe," Mah said, squatting down to point out a long crack snaking across all four lanes. "If we don't seal that, within five years we'll have big problems. Pieces coming out. Eventually the whole deck would collapse." Farther on, the cracks expanded into a fragmented jigsaw. "Look how bad this is. Very bad," he muttered.

Mah would later write all this up in a report, which would

hopefully lead to a Caltrans crew coming out to fill in the cracks. ("Nearly all departments of transportation are understaffed," said Sutter. "Their ability to identify problems is much better than their ability to solve them.") With proper maintenance, said Mah, the bridge should last another thirty or forty years, but no more. "Sooner or later it will need to be replaced," he said. "Material never lasts forever."

That's a fact the United States is learning the hard way. The most recent report on America's infrastructure by the American Society of Civil Engineers gives the nation's roads a grade of D. One-fifth of America's highways and one-third of its urban roads are in "poor" condition, inflicting $112 billion worth of extra repair and operating costs on American drivers.[25] According to the Federal Highway Administration, nearly one-quarter of all America's bridges are structurally deficient or functionally obsolete.

How bad can bad roads get? Afghanistan provides an extreme but relevant example. According to *The Washington Post*, the United States and other Western governments have poured more than $4 billion into building thousands of miles of new roads in that immiserated nation since 2001. Those roads are now in tatters, riddled with giant holes and crumbling pavement. Of course, some of the damage was caused by bomb blasts; but much of it is simply because after they were built, the roads got virtually zero maintenance.[26]

The state of America's 84,000-odd dams, most of the biggest of which are made with sand-based concrete, is even more unnerving. Their average age is fifty-six years, meaning quite a few are much older. Many were built to specifications far less stringent than those in force today, and so could break under the strain of a flood or earthquake. The American Society of Civil Engineers estimated that as of 2015 some 15,500 dams should be considered "high hazard potential"—meaning a failure would cause deaths. Bringing them up to current standards would cost tens of billions of dollars. Despite this, they don't get a lot of attention from overstretched state inspectors. Nationwide, there's only one safety inspector for every 205 dams. As of 2013, according to ASCE, South Carolina had only two people monitoring all of its 2,380 dams, and one of them was part-time.[27] So it was as unsurprising as it was tragic when in 2015 heavy rain collapsed 36 of the state's dams. As many as nineteen people were killed in the resultant flooding, according to *The New York Times*.[28] Scores of other dams around the country have failed since 2010. All told, hundreds of Americans are killed or injured each year due to the failure of the nation's sand-based roads, bridges, and dams.

Things are far worse in many developing nations, where building standards are low and the regulations that do exist are often ignored. A major Turkish developer told a newspaper a few years ago that during a building boom in the 1970s he routinely used unwashed sea sand to make concrete for buildings in Istanbul and elsewhere. Unwashed marine grains are cheaper to buy, but they are coated with salt that dangerously corrodes rebar. Concrete buildings made with sea sand pancaked by the dozens in Haiti's 2010 earthquake, and in 2013, Chinese officials halted construction of more than a dozen skyscrapers in Shenzhen that were found to contain unwashed sea sand.

Shoddy concrete was also likely a key reason for the disintegration of several buildings in a 1999 earthquake in Turkey and the collapse of an eight-story factory in Bangladesh in 2013 that killed more than 1,000 people. According to *The Financial Times*,[29] as much as 30 percent of Chinese cement is so low-grade that it produces dangerously flimsy structures known as "tofu buildings."

Cheaply made concrete is one of the reasons so many schools collapsed in China's 2008 Sichuan earthquake, killing thousands.

Vaclav Smil estimates that worldwide, as much as 100 billion tons of poorly manufactured concrete—buildings, roads, bridges, dams, everything—may need to be replaced in the coming decades. That will take trillions of dollars, and billions of tons of new sand.[30]

"Almost all the concrete structures you see today are doomed to a limited life span," writes Robert Courland. "Hardly any of the concrete structures that now exist are capable of enduring two centuries, and many will begin disintegrating after fifty years. In short, we have built a disposable world using a short-lived material, the manufacture of which generates millions of tons of greenhouse gases. Most of the concrete structures built at the beginning of the twentieth century have begun falling apart, and most will be, or already have been, demolished."[31]

We have built our world out of sand in the form of concrete—and it is starting to crumble.

CHAPTER 11

Beyond Sand

Armies of sand have built our cities, paved our roads, shown us distant stars and subatomic particles, spawned the Internet, and made our way of life possible. But extracting and deploying them on the immense scale of the twenty-first century has also brought destruction and death.

Since 2014, scores of people around the world have died, and many others have been injured, in accidents connected to sand mining[1]—run over by sand trucks, drowned in pits left by miners, or buried alive in sand avalanches. Most of them were children. Hundreds, likely thousands, more were driven from their homes by floods or river bank collapses brought on by sand mining, or threatened, assaulted, and injured while trying to stop illegal sand mining.

At least ninety-three people were reportedly murdered in violence related to illegal sand mining over the same period. The victims include an eighty-one-year-old teacher and a twenty-two-year-old activist who were separately hacked to death, a journalist burned to death, three police officers run over by sand trucks and another who had his throat slit and fingers chopped off, all in India.

In Kenya, a police officer was slashed to death with machetes, two truck drivers were burned alive, and at least half a dozen other people were killed in fighting over sand.

And all the while, more than 100 billion tons of sand and gravel[2] were ripped, scraped, and sucked up from floodplains, riverbeds, beaches, and the ocean floor, damaging rivers and deltas, killing coral reefs and fish, and bankrupting people who depend on those resources. Not to mention the damage caused by the other industries that put all that sand to work: the concrete makers, the land builders, the frackers.

All of that is just what I know about, from my own reporting and from tracking local media. There is no official tally of sand mining damage. There's no telling how much more is not reported or is deliberately kept out of the media.

So what is to be done?

Stronger government regulations can prevent, or at least mitigate, much of the harm caused by sand mining. They do in most of the developed world. Most restrictions on sand mining are relatively recent, however. Europe only got serious about regulations in the 1950s, after some of Italy's northern rivers were badly damaged by aggregate mining to build the highway network. France, the Netherlands, the United Kingdom, Germany, and Switzerland have banned river sand mining completely.[3] New York State passed its first laws regulating sand mining only in 1975. "Before that, it was up to municipalities or no one at all," said Bill Fonda, spokesperson for the New York State Department of Environmental Conservation.

Of course, there are plenty of questions about whether existing regulations adequately address sand mining, especially for frac sand. And sometimes those rules are simply ignored. Remember the $42 million fine that Hanson Aggregates had to pay to settle charges that it had stolen millions of tons of sand from San Francisco Bay?

Still, there are lots of safeguards in the system. In much of the United States, more than a dozen county, state, and federal government agencies have a say in determining who can mine sand where and under what circumstances. Mining companies are also generally required to restore the land, to a certain extent, after they're finished. (C. Howard Nye, the CEO of Martin Marietta, one of America's biggest construction aggregate companies, denounced all this regulation as "excessive" in testimony he gave to Congress in 2017.)[4]

Some agencies are waking up to the larger importance of sand. In 2011, Washington state authorities blew up a century-old dam because it was starving downstream ocean beaches of sand needed for clam habitat. The clams had all but disappeared. Now they are returning.[5]

Activism can make a big difference, too. Aggrieved citizens living near existing or proposed mines can and do lobby to keep them smaller, quieter, cleaner, and safer—or to keep them out of their backyard altogether. There are at least two proposed sand mining sites within an hour's drive of my home in Los Angeles that locals, unabashedly concerned about their views and property values as well as environmental impacts, have for years prevented from opening.

All of us have to recognize, though, that there is a price to be paid for protecting the environment and local residents' aesthetic sensibilities. If you forbid sand mining in your backyard—as many American communities do—then the sand to build your highways and shopping malls will have to be brought in from somewhere else. There still has to be a mine, somewhere. "It's like a garbage dump or a prison," said Ron Summers, former chair of the National

Stone, Sand, and Gravel Association. "Everyone wants one, but no one wants one near them."

In some situations, well-intentioned efforts to protect the local environment end up simply exporting the damage to somewhere with looser laws and less privileged citizens. In California's San Diego County in the early 1990s, federal, state, and local government officials cracked down on miners pulling sand out of the San Luis Rey River, after it became clear that all the digging was despoiling the river. Most of the mines soon shut down. With their local sources gone, San Diego concrete makers turned to importing sand from the nearby Mexican state of Baja California. That sparked a surge in mining in Baja—both legal and otherwise—that ravaged riverbeds, created a shortage of sand for local construction, and sparked street protests by villagers who blamed the mining for causing respiratory problems in their children. In 2003, Mexican officials resorted to temporarily banning exports of sand to California. Tempers have calmed down somewhat since, but the local press reports that illegal sand mining continues.[6]

Similarly, environmental concerns in North America and Europe are pushing sand mines ever farther from populated areas. Ironically, that is creating new environmental hazards.

The San Francisco Bay Area used to get much of its construction aggregate from the Livermore Valley, the place Henry Kaiser started mining. But the area gradually ran low on sand, and filled up with buildings that got in the way of mining. Aggregate miners found a new source north of the city, in the picturesque Russian River Valley in nearby Sonoma County. But as that area evolved from a rural backwater to a hub of wineries, organic farms, and outdoor tourism, the locals no longer wanted their landscape pocked with gravel pits or their roads filled with noisy trucks. So county supervisors banned mining along the river, forcing San Francisco to haul its sand from ever farther afield.

The same process is happening all over California and in many other places. The distances sand is hauled are increasing as quarries close to the big cities become depleted or are forced to close. About 80 percent of aggregates are hauled by truck; the rest goes by rail or barge. California officials estimate that if the average hauling distance for sand and gravel increases from twenty-five miles to fifty, trucks will burn through nearly 50 million more gallons of diesel fuel every year in the state alone, spewing more than half a million additional tons of carbon dioxide into the atmosphere.[7] Not to mention all the extra traffic and wear and tear on highways.

Pushing sand mines farther away also incurs financial costs. Sand is tremendously heavy, which makes it expensive to transport. The price of sand rises rapidly with each mile it travels. The increase in haul distances is one reason the inflation-adjusted price of construction sand in the United States has more than quintupled since 1978.[8] In major urban areas like San Francisco and Los Angeles, the price of trucked-in aggregate has risen so high that it now makes economic sense for developers to import some 3 million tons of sand and gravel by boat every year from a mine in Canada, almost 1,000 miles away.

The cost of aggregate is similarly being driven up internationally. "As efforts to curb illegal mining activities have been largely unsuccessful, sand and gravel reserves in many countries are expected to be depleted at a rapid pace through 2019. This will result in price hikes, especially in urban centers," noted a 2016 report by the Freedonia Group, an Ohio-based business research outfit. Developers in the Indian state of Telangana were forced to put several construction projects on hold in 2015 because a shortage of sand caused prices there to triple. A crackdown on illegal sand mining in Vietnam in early 2017 similarly sent prices skyrocketing. Worldwide, the average cost of a ton of construction sand has gone up nearly 50 percent in the last ten years, according to Freedonia's research.[9] That in turn makes concrete more expensive, which helps explain why housing prices have gone up so much in so many cities in the last couple of decades.

That might just be the beginning of the impact of dwindling sand supplies on the world economy. One key reason everything is made of concrete is that it's relatively cheap. If the cost of making a new building or road were to spike, it could hit regional and even national economies like an oil shock. In places like India where there are already severe housing shortages, a concrete price hike would only exacerbate the grim divide between those with means, who get to live in stable, waterproof structures, and the millions of others who have to make do in slums.

Tightening supplies are turning sand into more of a global commodity. Some $10 billion worth[10] of construction aggregate is sold across borders each year. It's one of North Korea's few exports.[11] Canadian sand is being barged even farther than California, all the way to Hawaii, where rules protecting beaches and inland sand dunes have cut off local supplies. Parts of Germany are so starved for sand that contractors import it from Denmark and Norway. In India, restrictions on sand mining have forced developers to import sand from Indonesia, the Philippines, and even the nation's archrival, Pakistan.

Things got especially weird in the Caribbean island nation of St. Vincent and the Grenadines in the 1990s. Alarmed by how thoroughly their own beaches were being plundered by the local construction industry, in December 1994 the little country banned beach sand mining and decreed that beginning the following year, all construction sand would have to be imported from nearby Guyana. Contractors, home builders, and truckers panicked, figuring prices would skyrocket. The result was an orgy of sand hoarding. Heavy machines dug away at island beaches around the clock, right through Christmas and New Year's Day. So much sand was stockpiled that it turned out to be more than anyone could use; the piles gradually blew away, the waves of grains clotting up roads and drainage pipes.[12] The ban was lifted and mining resumed. Many of the archipelago's dunes and beaches have since been decimated.

A key problem in much of the developing world is that all the regulations under the sun won't make any difference if no one enforces them. "There are very good laws on the books, but they are not applied," said Marc Goichot, a water issues researcher with the World Wildlife Fund. "The demand is too great and the ability of governments to enforce the laws is too low."

Which brings us to the issue of corruption. Bribes and payoffs, officials on the take, are probably the main reason illegal sand mining continues on such a massive scale—and why, at the time this book is being written, Paleram Chauhan's killers still haven't been brought to trial.

It's a worldwide problem. Corruption in the aggregate business—as in most extractive industries—runs the gamut from villagers slipping the local magistrate a few banknotes to turn a blind eye to an illegal pit, to employees of giant multinationals participating in major-league villainy. In 2010, two French nationals working in Algeria for Lafarge, one of the world's biggest cement and aggregate firms, fled the country one step ahead of police who were after them on money laundering and corruption charges.[13] The same company admitted in 2016 that its Syrian subsidiary had paid off armed groups, possibly including ISIS, to leave one of its cement plants alone; their CEO resigned in the ensuing scandal.[14]

In some places, illegal sand miners have the added protection of powerful people who are involved in the industry. According to Global Witness, a British research activist group, two extremely wealthy members of Cambodia's senate[15] run many of that nation's sand mines. Members of national and provincial governments of India and Sri Lanka also reportedly have their hands in the trade.

It's often local officials, the very people who are supposed to be protecting the interests of their communities, who are the worst offenders. In 2015, on the Indonesian island of East Java, two farmers—Salim Kancil, fifty-two, and Tosan, fifty-one (many Indonesians use only one name)—led a series of protests against an illegal beach sand mining operation. The mine operators threatened to kill them if they kept interfering; the farmers reported the threats to the police and asked for protection. Soon after, at least a dozen men attacked Tosan, ran him over with a motorcycle, and left him for dead in the middle of the road. Then they moved on to Salim's house. They beat him and hauled him off to the village hall, where he was battered with clubs and stones and finally stabbed to death. His body was left on the street with his hands tied behind his back.

Police eventually arrested thirty-five people. Two of them were sentenced to twenty years in prison for masterminding the attack—both of them local officials. One was the village chief.

(The sand industry really seems to attract some of the worst people in all of Indonesia. There's also Chep Hernawan, an Indonesian businessman involved in real estate, plastics recycling, and sand mining. The founder of an organization dedicated to imposing Islamic law on Indonesia, he is also a vocal supporter of jihadi terrorists. He offered to donate land for the burial of three men executed for their role in the 2002 Bali nightclub bombings, and in

2015 he told CNN that he had paid the travel expenses of 156 of his countrymen who went to Iraq and Syria to fight with ISIS.)[16]

A few months before the East Java murders, I made my way to a sand mine on the neighboring island of Bali, far inland from the tourist beaches. It looked like Shangri-la after a meteor strike. Smack in the middle of a beautiful valley winding between verdant mountains, surrounded by jungle and rice paddies, was a raggedy fourteen-acre black pit of exposed sand and rock. On its floor, men in shorts and flip-flops swung sledgehammers at rocks and hoisted shovelfuls of sand and gravel into clattering, smoke-belching sorting machines.

I wandered around the place for a couple of hours, trying to find out who was in charge. No one seemed to know—at least, no one was willing to give any names to a foreign journalist. What are the odds this mine was operating legally? "Seventy percent of the sand miners have no permits," Nyoman Sadra, a former member of the regional legislature, told me later. As an article in *The New York Times Magazine* recently put it, "the sand trade is . . . sustained by a devilishly inbuilt chain of plausible deniability. . . . Sand mining is executed by an endless array of small, independent, often temporary players, largely working at night and in secret. And each step of the line of production is separated from the rest: The sand moves from diggers to truckers to dealers to builders with each link in the chain knowing as little as possible about where the sand they're buying comes from or who mines it—for obvious reasons, they don't want to know."[17]

It just takes a few handfuls of strategically distributed cash to get the local police to leave the sand miners alone. Even companies with permits spread money around so they can get away with digging pits wider or deeper than they're supposed to. "They just bribe government officials," Suriadi Darmoko, an activist with the Indonesian Forum for the Environment, said. "It's an open secret."

The village chief convicted of Salim Kancil's murder, for one, admitted to paying off police officers to protect the mine.

I got a good look at how this plays out on the ground while I was in India. I spent several days there with Sumaira Abdulali, India's foremost campaigner against illegal sand mining. Abdulali is a decorous, well-heeled member of the Mumbai bourgeoisie, gentle of voice and genteel of manner. For years she has been traveling to remote areas in a leather-upholstered, chauffeur-driven sedan, snapping pictures of sand mafias at work. In the process she's been insulted, threatened, pelted with rocks, pursued at high speeds, had her car windows smashed, and been punched in the mouth hard enough to break a tooth.

Abdulali got involved when sand miners started tearing up a beach near Mumbai where her family has vacationed for generations. In 2004 she filed the first citizen-initiated court action against sand mining in India. It made the newspapers, which in turn brought Abdulali a flood of calls from others around the country who wanted her help stopping their own local sand mafias. Abdulali has since helped dozens file their own court cases and keeps a steady stream of her own well-documented complaints flowing to local officials and newspapers. "We can't stop construction. We don't want to halt development," she says in British Indian–accented English. "But we want to put in accountability."

Abdulali took me to the rural town of Mahad, on India's western coast, where sand miners once smashed up her car. Sand mining is completely banned in the area because of its proximity to a protected coastal zone. Nonetheless, in the jungle-draped hills not far outside town, we came to a gray-green river on which boats, in plain view, were sucking up sand from the river bottom with diesel-powered pumps. The riverbanks were dotted with huge piles of grains, which men driving excavators were shoveling onto trucks.

Soon after, back on a main road, we found ourselves behind a small convoy of three sand trucks. They rumbled, unmolested, past a police van parked on the side of the road. A couple of cops idled next to it, watching the traffic going by. Another was inside the van taking a nap, his seat fully reclined.

This was too much for Abdulali. We pulled up alongside the van. An officer who appeared to be in charge was lounging inside, wearing a khaki uniform with stars on his shoulders and black socks on his feet. He had taken his shoes off.

"Didn't you see those trucks carrying sand that just went past?" Abdulali asked.

"We filed some cases this morning," answered the cop genially. "We're on our lunch break now."

As we drove away, we passed another truck filled with black market sand, parked just a few hundred yards down the road.

Some time later I told a local government official about this encounter. He wasn't surprised. "The police are hand in glove with the miners," said the official, who asked not to be named. "When I call the police to escort me on a raid, they tip off the miners that we are coming." Even in the cases he'd brought to court, no one was convicted. "They always get off on some technicality."

It's clear citizens can't rely only on governments to enforce the laws that should control sand mining. Another way to approach the problem could be via collective consumer action, following the model of the fair-trade movement. There are many international programs that certify whether your coffee, diamond ring, or wooden table was created without causing undue environmental damage or exploiting workers or funding warlords. None of them is a complete or foolproof solution, of course, but they're better than nothing. Why not set up a similar independent consumer watchdog for the sand industry?

Technology might also offer some help. There are many researchers and scientists around the world who are working on ways to make concrete that lasts longer, which would cut down on the amount of sand needed every year.

One of concrete's key shortcomings is its vulnerability to cracks through which moisture seeps in, corroding the rebar inside. What if the concrete could just fill in those cracks all by itself? It turns out self-healing concrete is actually possible. Researchers in Europe are working with bacteria that excrete the mineral calcite and can also survive dormant for decades encased in concrete. When a crack forms, the encroaching water wakes up the bacteria, which starts to excrete calcite, filling up the crack. The process works just fine in the lab and is under development for real-world use.[18]

Another approach is to embed hydrogels, polymers that expand as they absorb moisture (they're used in baby diapers, among other products); when water seeps into a crack, the hydrogel expands, filling it in. Scientists in South Korea are also experimenting with a protective coating containing microcapsules full of a solution that turns solid on exposure to sunlight. In theory, a crack in the concrete would break open the capsules, releasing the solution, which would turn solid in the sunlight. There are several other methods of getting things to automatically ooze into cracks being researched in labs in several countries.

There's also what's called geopolymer concrete, which replaces cement with a binding agent made of natural materials and industrial by-products like fly ash, a powdery leftover from burning coal in power plants. Cement is the component of concrete that requires by far the most energy to produce, and the production of which generates even more greenhouse gases as a waste product, so removing it from the mix would be a huge help to the atmosphere. Versions of this geopolymer concrete are already being used in a few places around the world, mostly as pavement. Other researchers are also looking into a range of additional ways to reduce emissions from cement making.

Since the steel rebar is the component most likely to fail in reinforced concrete, what about replacing it with something more dependable? A Norwegian company is marketing bars made of basalt fibers that it touts as a corrosion-proof alternative to steel rebar. Other researchers are trying to replace rebar with woven strips of carbonized bamboo. Concrete reinforced with fiberglass is also stronger and longer lasting, though it's not in widespread use. Meanwhile, a Danish company claims to have developed a technique for using desert sand to make concrete, though it has yet to bring it to market.

All of these ideas sounds great in principle. Whether they can be made to work at a reasonable price in the real world is a question as yet unanswered.

What about recycling sand after it has been turned into something else? It can be done, but only on a relatively small scale. Glass can be effectively recycled (a New Zealand beer company has even developed a machine to convert bottles directly into beach-ready sand), but the glass industry makes up just a fraction of overall sand use. Most sand goes into making concrete. It is possible to crush up and reuse concrete, but it's not cheap—it requires removing the rebar, for one thing—and recycled concrete is considered good enough to use only for low-quality applications like road base and sidewalks. The market for recycled concrete is growing, but it's still just a tiny slice of the pie. Asphalt is much easier to recycle, and about 73 million tons of it is reused each year.[19] But again, that's a relative drop in the bucket.

In any case, buildings and roads aren't bottles. They're not meant to be used once and then tossed. They are meant to be used continuously for decades. They stay put. The sand that goes into them is frozen in place, taken out of circulation perhaps forever.

It is possible to make more sand, but it's not easy or cheap. Crushing rock or pulverizing concrete down to small grains can work. Japan, for one, has relied heavily on such man-made sand since it banned marine dredging for construction sand in 1990.[20] But making artificial sand is more expensive than harvesting the natural kind, and the resulting grains are ill suited for many applications; the freshly shattered grains are often too angular, among other shortcomings. We can dredge up some of the sand that's trapped behind dams, but that's also costly.

We can use alternative substances for some purposes. Fly ash, copper slag, and quarry dust, for instance, can replace the sand in some kinds of concrete. In India, there's a project under way to use shredded plastic trash instead of sand to make concrete, which offers the opportunity to reduce both the amount of sand taken from riverbeds and the amount of trash that goes into landfills. In Australia, an engineer is working on a method to make pavement from a combination of coffee grounds and waste products from steel production.

All of these efforts can and hopefully will help. But the sheer volume we need to build our cities makes it all but impossible to replace aggregate on a large scale. What other substance can we possibly find 50 billion tons of, every year?

Ultimately, there's only one long-term solution: human beings have to start using less sand. For that matter, we have to start using less of *everything*.

You've heard it before. Human beings are eating up the planet. We're living way beyond our environmental means. We're burning too much oil, catching too many fish, cutting down too many trees, pumping too much freshwater. We're using too much *phosphorus*, for God's sake; it's a crucial ingredient in crop fertilizer that comes only from certain types of rock, and supplies of those rocks are running low.[21]

We're even running out of commodities you've never heard of but rely on every day. Today's high-tech gadgets, from smartphones to solar panels, use a bevy of rare, obscure metals like tantalum and dysprosium. There are very few sources for most of those things, and supplies are getting alarmingly tight, as David S. Abraham details in his book *The Elements of Power*. "At no point in human history have we used more elements, in more combinations," writes Abraham. "The future of our high-tech goods may lie not in the limitations of our minds, but in our ability to secure the ingredients to produce them . . . our ingenuity will soon outpace our material supplies."[22]

The amount of raw material—the sheer tonnage of iron, wood, oil, sand, and all other substances—used by human beings has more than tripled just since 1970.[23] The World Wildlife Fund calculates that humans have been using up natural resources faster than nature can replenish them for forty years now—that is, we're cutting trees down faster than new ones can mature, harvesting fish faster than new stocks can be bred, and so on. The same, of course, goes for sand. New sand is constantly being created as the elements erode mountains, but the amount we use far exceeds the amount being made. It would require one and a half Earths to sustainably generate all the materials we use each year.[24] If everyone on Earth had an American standard of living, we'd need *four* and a half Earths.[25]

In the island nation of Cape Verde, ironically, overconsumption of other resources is forcing people to turn to sand mining. The 2013 documentary *Sandgrains* tells the story of a village where families have turned to digging up sand by the bucketload from the ocean floor and selling it, because industrial-scale fishing has decimated the marine life they used to depend on.[26] The villagers still rely on the sea to survive, but now they take its sand instead of its fish.

Consumption of almost every important resource—everything from wheat to paper to copper—is headed only one way: up.[27] The size of typical new American houses has increased by more than 1,000 square feet since 1973, according to the US Census Bureau, to an all-time high of 2,679 square feet. At the same time, the number of people living in those houses has declined from an average of 3 to 2.5. Combined, those figures mean that the amount of living space the average American takes up has nearly doubled in the last forty years.[28] Think about how much wood, wiring, energy, and sand went into making all those extra rooms.

The Western world invented the modern good life, with its car-dependent suburbs, oversize houses, SUVs, and TVs in every room. It is not physically possible to replicate that lifestyle worldwide. Already, according to a recent study by Austria's Alpen-Adria-Universität, fully industrialized countries of the West use up one-third of all global resources, and more than half of all fossil fuels and industrial minerals, including sand. Nonetheless, resource consumption in China, India, and many other countries is catching up fast.

How could it not? Economic growth is raising standards of living all over the developing world. Since 1990, nearly a billion people have been lifted out of extreme poverty, and 1.2 billion have risen into the global consuming class—people with money to spend on things beyond daily necessities. In the coming decades, as many as 3 billion are projected to rise into the global middle class.[29]

Meanwhile, at the other end of the spectrum, some 1.6 billion people around the world live in inadequate shelters, the United Nations estimates.[30] More than 100 million have no homes at all. Providing a decent place to live for those people will require a gargantuan use of resources. By 2030 the world will need to add 4,000 new affordable housing units every *hour* to meet the demand. India alone will need housing and urban infrastructure for more than 400 million people by 2050. That's more than the entire population of the United States.

Sooner or later, all of this will inevitably lead to shortages of sand. In fact, that's already happening. A 2012 report by California's Department of Conservation warns that the state has access to only about one-third of the sand and gravel it will require over the next fifty years. The United Kingdom has increasingly turned to the seas as its land-based sand mines have come under pressure; ocean-floor sand now provides about one-fifth of the nation's needs. But those supplies are predicted to last only another fifty years.[31] Vietnam's Ministry of Construction warned in 2017 that the country was on track to run out of sand completely in less than fifteen years.

The very structures we've made out of sand are now getting in the way of our getting more. "Our high-quality aggregate is getting covered with shopping centers," said Larry Sutter, the concrete expert at Michigan Technological University.

Of course, there is still a lot of sand on the planet. We're not going to literally use it all up. We won't have tribes of biker mutants battling each other for the last truckloads of the stuff any time soon. But the sand situation is in many ways comparable to that of other crucial natural resources. There is plenty on the planet—but it's often found a long way from where the people who need it live, or it can be extracted only at the risk of severe environmental damage.

Consider what's happening with fossil fuels. There's still plenty of oil and natural gas left in the ground. But a lot of the easily accessible hydrocarbons close to the surface are gone. That has forced the energy industry to turn to fracking and to subsea fields like the one so disastrously tapped by the Deepwater Horizon, the BP rig that exploded in the Gulf of Mexico in 2010. In other words, we can still get all the fossil fuels we need, but at an ever-growing environmental and social cost.

Or think about freshwater, which is in frighteningly short supply from the Middle East to the American Southwest. Worldwide, there is plenty. But getting water from somewhere that has lots, like Canada, to somewhere that has little, like Jordan, would be a tremendously expensive proposition—and that's assuming Canada would be willing to part with it. As the battles over beach sand in southern Florida show, even neighboring counties can be selfish when it comes to sharing their sand.

How much nastier could this get? Would countries with surplus sand hoard it at the expense of their sand-starved neighbors? Yes, they would. In 2007 China did exactly that, temporarily suspending exports of construction sand to Taiwan. In 2009 Saudi Arabia did the same, briefly banning sales of construction sand to other gulf countries because of domestic shortages.

You read that right: Saudi Arabia is worried about running out of sand.[32]

Meanwhile, the armies of sand that we continue to mobilize are abetting what may soon be everyone's biggest worry: climate change. Transforming sand into concrete and glass requires energy—enormous amounts of it from power plants fired by coal and natural gas. More important, sand is the symbiotic partner of fossil fuels, the unsung but essential partner of the oil and gas industry. Sand makes the roads that make gasoline- and diesel-burning automobiles useful. Sand makes the suburbs and shopping malls and office parks that make automobiles indispensable. Sand makes it possible to unlock billions of barrels of once-inaccessible oil and natural gas.

It's easy to wax self-righteous about corporations ravaging the natural world. But when it comes to some natural resources—prominently including oil and sand—all of us need the things those corporations produce. Aggregate industry professionals like to gripe about LICAs—"low information community activists"—and CAVEs, "citizens against virtually everything." To a certain extent, they have a point. No one who has grown up with the comforts and conveniences of modern life really wants to give them up. Without oil and gas, we have no cars and trucks, and much less energy (at least until wind and solar ramp up). Without sand, we have no modern cities, no modern life. It's flat-out impossible to extract those resources from the reluctant earth without inflicting *some* damage, without making some changes to the natural world. It's dishonest or naive to pretend even a fraction of the 7 billion of us can have any sort of reasonable standard of living without doing *any* harm to the planet. So the question really is, how far are we willing to go? How much damage are we willing to do, and where, and to what?

Whenever someone says that population growth is putting the world in danger of running out of some critical natural resource, optimists (and self-interested industrialists) usually respond by pointing out that people have been warning about exactly this scenario since the days of Thomas Malthus in 1798—and it still hasn't happened. Technological breakthroughs, policy adjustments, and new discoveries have always carried us through predicted crises, from the ozone hole to peak oil.

That's true. But it won't necessarily always be true.

Many of the disasters we've been warned about were avoided *because* we were warned about them and took action to prevent them. The ozone layer didn't magically start replenishing itself. It has been replenished because the nations of the world recognized that ozone holes were a huge problem, and agreed to stop using chlorofluorocarbons and other gases that were gouging out those holes.

It's also crucial to bear this in mind: The speed and scale of change in today's world is utterly without precedent. It's miles beyond anything ever seen in 4 million years of human history. "Britain took 154 years to double economic output per person, and it did so with a population (at the start) of nine million people," write the authors of *No Ordinary Disruption*, a recent report on world economic trends by the McKinsey Global Institute. "The United States achieved the same feat in fifty-three years, with a population (at the start) of ten million people. China and India have done it in only twelve and sixteen years, respectively, each with about 100 times as many people. In other words, this economic acceleration is roughly 10 times faster than the one triggered by Britain's Industrial Revolution and is 300 times the scale—an economic force that is 3,000 times as large."[33] Economic growth across the developing world, they add, means that by 2025, the consumer class—those with enough extra income to buy nonessential items—will grow to a total of 4.2 billion consumers. Fifty years ago, there weren't even that many people on the planet, let alone that many shopping for smartphones.

Our way of life worked in the last century because the number of people living it—almost exclusively in Western countries—was relatively small. Most of the world's people were poor. For the first time in history, that is changing. Western industrialized nations are still consuming just as much, and now everyone else is starting to consume more as they move up the economic ladder.

Those new consumers want the same car and gadget-enabled life we enjoy in the West. And they're getting it. In 1995, only 7 percent of Chinese city dwellers owned refrigerators. Twelve years later, 95 percent did. All this rapid growth, warned the US National Intelligence Council in a 2012 report, "will mean a scramble for raw materials and manufactured goods."[34] From fossil fuels to food, minerals, timber, you name it, "the scope and size of resource consumption, and the associated environmental impacts, risk overwhelming the ability of states, markets and technology to adapt," declared a 2012 report[35] from Chatham House, a venerable British think tank.

Sand is just one aspect, one element of the much larger problem of overconsumption. Remember, quartz sand is perhaps the most abundant substance on the planet's surface. If we're running out of *that*, we really need to rethink how we're using everything.

Don't get me wrong. I like my single-family home with its capacious refrigerator, big-screen TV, central air-conditioning, and flock of laptops, tablets, and cell phones as much as anyone. I'm not suggesting we cast off all our material goods and go live in the woods. But I have spent enough time in more modest circumstances to know that we can live a perfectly comfortable, thoroughly modern life in a smaller house with fewer appliances and fewer cars and less stuff in general than is the norm in twenty-first-century America.

One promising development in this direction is the rise of the "sharing economy," a term that must have been invented by some marketing rep at Uber or Airbnb or one of the many other new outfits that make it easy to *rent* surplus resources. (I'll call it "sharing" when they stop charging.) Semantic quibbles aside, these services represent a novel and overdue way to cut down on the enormous waste of postindustrial economies. Among other things, they could help reduce our consumption of sand.

In America, most adults own cars. And most of those cars spend most of their time sitting still, parked. Ride-hailing services are making it easier than ever for city dwellers, at least, to get rid of their own cars and pay for rides only when they need them.

How might reducing car ownership save sand? Today, the typical American home is built with a garage and a driveway—car-support structures that are made with concrete, which is to say, with sand. If you didn't own a car, however, you wouldn't need those structures. The amount of sand required to build your house would be reduced by many tons.

Similarly, if Airbnb et al mean we can stay in people's extra rooms when traveling instead of in a hotel, fewer hotels need to be built. The legions of sand that would have been drafted to build those hotels, with their driveways and parking lots, could instead be left in the ground. (Not to mention all the other resources that would be saved.)

And if we have less need for new buildings, the expansion of cities might slow down. Then we wouldn't need to dredge up so much ocean sand to create artificial land. Maybe we'd also reduce water use enough that we could stop draining it from drylands, lessening the threat of desertification.

Making fewer cars and buildings also means we'd use less energy, reducing our need for fossil fuels. Then there would be less need for fracking, which would mean we could stop tearing up Wisconsin farmland to get frac sand.

The sands of time are running out. Our houses are built upon sand. Pick your metaphor. But understand: it's not just a metaphor. Sand is the floor beneath our feet and the roof over our head. It is the substrate of modernity. On top of it we have built an economy and a society that depends on sand for far more purposes than Ernest Ransome, Michael Owens, and even Dwight D. Eisenhower could have dreamed of.

And yet, sand is about the most taken-for-granted natural resource in the world. Hardly anyone thinks about it—where it comes from or what we do to get it. But in a world of 7 billion people, more and more of whom want apartments to live in, offices to work in, malls to shop in, and cell phones to communicate with, we can't afford that luxury anymore.

It once seemed like we had such boundless supplies of oil, water, trees, and land that we didn't need to worry about them. But of course we're learning the hard way that none of those things are infinite, and the price we've paid so far for using them is rising fast. We're having to learn to conserve, reuse, find alternatives for, and generally get smarter about how we use those natural resources. We have to start thinking that way about sand, too.

But we also need to understand that the bigger issue isn't just about being more careful or smarter about how we use individual resources. It's about how we use *all* those resources. It's about figuring out a way to build a life for 7 billion people on a foundation sturdier than sand.

ACKNOWLEDGMENTS

I could not have written this book without the many, many people in many places around the world who gave generously of their time, expertise, and encouragement. I am especially indebted to Aakash Chauhan, who risked his own safety to help me tell the story of the murder of his father, Paleram Chauhan, in an article for *Wired* magazine from which this book grew. Chauhan continues to speak out courageously against India's sand mafias, for which he deserves even more thanks. The indefatigable Sumaira Abdulali, perhaps India's foremost campaigner against illegal sand mining and other overlooked environmental scourges, was also a key ally on that first article. So was Vikrant Tongad, founder of Social Action for Forest and Environment, and journalist Kumar Sambhav. Speaking of journalists: my utmost respect to the many in India who regularly cover the violence and destruction sand miners are inflicting on their country—and who themselves are not infrequently targets of that violence. And a heartfelt hat-tip to Jakob Villioth, whom I have never met but whose report on the global sand industry for Ejolt.org first made me aware there even is such a thing.

In North Carolina, Alex Glover and David Biddix were excellent guides to the sights of Spruce Pine, and bountiful sources of information on its unique history and geology. Dr. Tom Gallo was kind enough to share not only his story but his expertise in the quartz industry. Jessica Roberts of Roskill Information Services also provided invaluable technical details.

ACKNOWLEDGMENTS

In Wisconsin, my thanks to Ken Schmitt and Donna Brogan for squiring me around their respective counties, Crispin Pierce for letting me tag along on a research trip with his students, and the Wisconsin Center for Investigative Journalism for their database on fracking sites.

In Florida, activists Dan Clark and Ed Tichenor, and Robert Weber, coastal coordinator for the town of Palm Beach, gave generously of their time to show me different aspects of beach nourishment.

In Dubai, much appreciation to journalist Jim Krane for introductions to some key locals (and for his excellent book), and to Lubna Sharief Takruri, fixer extraordinaire from the West Bank to the Persian Gulf.

In China, my thanks to Kong Lingyu, smartphone impresario and outstanding fixer/interpreter; to documentary makers Qiong Wang and Xiao Qiping, who took me to some obscure corners of Poyang Lake; and to David Shankman, America's unofficial ambassador to the city of Nanchang. Also to Jennifer Turner at the Wilson Center, who introduced me to a whole squad of contacts. Chief among them was Luan Dong, who arranged for me to give a BEER talk in Beijing, which was just as much fun as it sounds like.

In Indonesia, Anton Muhajir got me everywhere I needed to go. In Cambodia, Oudom Tat did likewise. I am also grateful to Alex Gonzalez-Davidson and his colleagues at Mother Nature Cambodia for their brave work and the help they gave me in Koh Kong. Thanks also to Jacob Kushner for reporting help from Kenya, and to Peter Klein for recommendations to interpreters/fixers in many countries.

The unsung statisticians at the United States Geological Service, tabulators of sand usage for over a century, deserve a special commendation, as does Sterling Kelly at the Bureau of Labor Statistics, who also provided me with the excellent Jorge Luis Borges quote that introduces Part I. Pascal Peduzzi of the United Nations Environment Programme, who authored the first authoritative report on the sand crisis, provided some key early material. Thanks to Bailey Wood of the National Stone, Sand, and Gravel Association for several rounds of answers and introductions. Geologist Michael Welland was a huge help with my early research, both by phone and via his excellent books; I was very sorry to hear of his passing while this book was still under way.

ACKNOWLEDGMENTS

Special thanks, of course, to Lisa Bankoff, literary agent nonpareil, and to Jake Morrissey, editor extraordinaire at Riverhead Books. Both of them gave critical advice in shaping this book, which I didn't always accept with the best of grace at the time, but most of which turned out to be exactly right. My thanks also to my editors at *Wired*, *The New York Times*, the *Guardian*, *Pacific Standard*, and *Mother Jones*, who published what would become portions of this book, especially *Wired*'s Adam Rogers, who shepherded my first big feature on the topic. I am very grateful to Tom Hundley and the rest of the staff at the Pulitzer Center on Crisis Reporting, who provided grants that helped make all the travel possible. Thanks also to Michelle Delgado for the cheeriest assistance with research and logistics any scrivener could ask for. And to Vinnie Hollywood, Vladimir Reptilio, and the whole crew at Blessed Reptile Productions, for all that they do.

I am also indebted to many colleagues, friends, and relatives. Taras Grescoe, Tom Zoellner, David Davis, Justin Pritchard, Linda Marsa, Scott Carney, Hector Tobar, and Cari Lynn gave kindly and insightful feedback on chapter drafts and/or shared their wisdom on the book-writing business. They're all top-notch writers, and you should buy their books. Special thanks to Adara and Isaiah Beiser Shilling for tolerating their dad's many trips away over the past couple of years (and for joining me on one of them), and to Kaile Shilling, who plowed through and constructively critiqued two full-length drafts, as if being married to me hadn't already obliged her to hear more than she ever wanted to about sand.

NOTES

In the course of researching this book, I have interviewed more than a hundred people and made my way through upwards of a thousand studies, reports, news articles, and other documents. Unless otherwise noted, all quotes are from interviews, in person or by phone. I have footnoted only those facts that I think might be particularly surprising, contentious, or difficult for the reader to verify on his/her own.

Chapter 1: The Most Important Solid Substance on Earth

1. "World Construction Aggregates," Freedonia Group, 2016.
2. Michael Welland, *Sand: The Never-Ending Story* (Berkeley: University of California Press, 2009), 1–2.
3. Welland, *Sand,* 240.
4. Tom's of Maine, *"Hydrated Silica,"* http://www.tomsofmaine.com/ingredients /overlay/hydrated-silica; American Dental Association, "Oral Health Topics -Toothpastes," http://www.ada.org/en/science-research/ada-seal-of-acceptance /product-category-information/toothpaste.
5. Pascal Peduzzi, "Sand, rarer than one thinks," *United Nations Environment Programme Report,* March 2014, 3.
6. "World Construction Aggregates," 2016.
7. United Nations Department of Economic and Social Affairs, "2018 Revision of World Urbanization Prospects." https://www.un.org/development/desa /publications/2018-revision-of-world-urbanization-prospects.html.
8. United Nations Department of Economic and Social Affairs, "World Urbanization Prospects," 2014.
9. Peduzzi, "Sand, rarer than one thinks," 1.
10. Ana Swanson, "How China used more cement in 3 years than the U.S. did in the entire 20th century." *Washington Post,* March 24, 2015. https://www

.washingtonpost.com/news/wonk/wp/2015/03/24/how-china-used-more
-cement-in-3-years-than-the-u-s-did-in-the-entire-20th-century/?utm
_term=.bbae0f4bc08a.

11. Welland, *Sand*, 252–53.

12. Peduzzi, "Sand, rarer than one thinks," 6.

13. Raymond Siever, *Sand* (New York: Scientific American Library), 1988, 17.

14. Welland, *Sand*, 16.

15. Mark Miodownik, *Stuff Matters: Exploring the Marvelous Materials That Shape Our Man-Made World* (Boston: Houghton Mifflin Harcourt, 2014), 140.

16. Welland, *Sand*, 1–23.

17. Siever, *Sand*, 55.

18. Thomas Dolley, "Sand and Gravel: Industrial," *US Geological Survey Mineral Commodity Summaries*, January 2016, 144–45.

19. "What Is Industrial Sand?" National Industrial Sand Association, http://www.sand.org/page/industrial_sand.

20. Welland, *Sand*, 13.

21. Jason Christopher Willett, "Sand and Gravel (Construction)," *US Geological Survey Mineral Commodity Summaries*, January 2017, 142.d/mining/usgs sand construct 2016.

22. "Annual Review 2015–2016," European Aggregates Association, 4.

23. "Specialty Sands," Cemex, http://www.cemexusa.com/ProductsServices /LapisSpecialtySands.aspx.

24. Denis Cuff, "State sued over sand mining in San Francisco Bay," *East Bay Times*, January 31, 2017.

25. Erwan Garel, Wendy Bonne, and M. B. Collins. "Offshore Sand and Gravel Mining," *Encyclopedia of Ocean Sciences*, 2nd ed., John Steele, Steve Thorpe, and Karl Turekian, eds. (New York: Academic Press, 2009), 4162–170.

26. "The Mineral Products Industry at a Glance," Mineral Products Association, 2016, 10.

27. Garel, et al., "Offshore Sand and Gravel Mining," 3.

28. G. Mathias Kondolf, et al., "Freshwater Gravel Mining and Dredging Issues," *White Paper Prepared for Washington Department of Fish and Wildlife*, April 4, 2002, 49, 64.

29. Peduzzi, "Sand, rarer than one thinks," 4.

30. Global Witness, "Shifting Sand," May 2010, 18.

31. Wildlife Conservation Society Cambodia, "Cambodia's Royal Turtle Facing Increased Threats to Survival," https://cambodia.wcs.org/About-Us/Latest -News/articleType/ArticleView/articleId/8888/Cambodias-Royal-Turtle -Facing-Increased-Threats-to-Survival.aspx.

32. Kondolf, et al., "Freshwater Gravel Mining and Dredging Issues," 71, 81–88.

33. Felicity James, "NT sand mining destroying environmentally significant area without impact assessment, EPA confirms," ABC News, November 1, 2015; http://www.abc.net.au/news/2015-11-01/no-environmental-assessment- of-nt-sand-mining/6901840.

34. Kiran Pereira, "Curbing Illegal Sand Mining in Sri Lanka," *Water Integrity in Action* report, 2013, 14–15.

35. Supreme Court of India, *Deepak Kumar and Others v. State of Haryana and Others*, 2012.

36. Kondolf, et al., "Freshwater Gravel Mining and Dredging Issues," 108.

37. Ibid., 60, 80.

38. D. Padmalal and K. Maya, *Sand Mining: Environmental Impacts and Selected Case Studies* (New York: Springer, 2014), 40, 60, and Kondolf, et al., "Freshwater Gravel Mining and Dredging Issues." 62, 65.

39. "Heavy Machinery Miyun Pirates . . . ," *The Beijing News*, December 21, 2015; http://epaper.bjnews.com.cn/html/2015-12/21/content_614577.htm?div=-1.

40. "Sand mining a trigger for crocodile attacks," *The Times of India*, March 15, 2017; http://timesofindia.indiatimes.com/city/kolhapur/sand-mining-a-trigger-for-croc-attacks/articleshow/57638419.cms.

41. "Attorney General Lockyer Files $200 Million Taxpayer Lawsuit Against Bay Area 'Sand Pirates,'" official press release, October 24, 2003; https://oag.ca.gov/news/press-releases/attorney-general-lockyer-files-200-million-taxpayer-lawsuit-against-bay-area.

42. Interview with Bill Fonda, New York State Department of Environmental Conservation, March 2, 2017.

43. Peduzzi, "Sand, rarer than one thinks," 7, and Orrin H. Pilkey and J. Andrew G. Cooper, *The Last Beach* (Durham, NC: Duke University Press, 2014), 32.

44. Cited in "A shore thing: An improbable global shortage: sand," *The Economist*, March 30, 2017; economist.com/news/finance-and-economics/21719797-thanks-booming-construction-activity-asia-sand-high-demand.

45. Many of the details about Paleram Chauhan's case come from my interviews with his family members and court documents they provided to me.

46. "Site visit to ascertain the factual position of illegal sand mining in Gautam Budh Nagar, Uttar Pradesh," official report, August 8, 2013.

Chapter 2: The Skeleton of Cities

1. "The San Francisco Earthquake, 1906," *EyeWitness to History*, www.eyewitnesstohistory.com (1997).

2. Robert Courland, *Concrete Planet: The Strange and Fascinating Story of the World's Most Common Man-Made Material* (Amherst, NY: Prometheus Books, 2011), Kindle Location 1881.

3. Michael Welland, *Sand: The Never-Ending Story* (Berkeley: University of California Press, 2009), 235.

4. Mark Miodownik, *Stuff Matters: Exploring the Marvelous Materials That Shape Our Man-Made World* (Boston: Houghton Mifflin Harcourt, 2014), 56.

5. Courland, *Concrete Planet*, Kindle Location 1009.

6. Courland, *Concrete Planet*, Kindle Locations 992–994.

7. Earl Swift, *The Big Roads: The Untold Story of the Engineers, Visionaries, and Trailblazers Who Created the American Superhighways* (Boston: Houghton Mifflin Harcourt, 2011), Kindle Location, 85.

8. Courland, *Concrete Planet*, Kindle Locations 1248–1252, 1383, 1421.
9. Miodownik, *Stuff Matters*, 58.
10. Ibid., 59.
11. "Cement Manufacturing Basics," Lehigh Hanson, http://www.lehighhanson.com/learn/articles.
12. Courland, *Concrete Planet*, 2033–2089, 2157, 2325.
13. "The Thames Tunnel," Brunel Museum, http://www.brunel-museum.org.uk/history/the-thames-tunnel/.
14. Vaclav Smil, *Making the Modern World: Materials and Dematerialization* (Hoboken, NJ: Wiley, 2013), 28.
15. Courland, *Concrete Planet*, 2755.
16. "Cement Statistical Compendium," US Geological Survey, https://minerals.usgs.gov/minerals/pubs/commodity/cement/stat/.
17. Courland, *Concrete Planet*, 3005–3008.
18. Miodownik, *Stuff Matters*, 61.
19. Courland, *Concrete Planet*, 3112, and Miodownik, *Stuff Matters*, 60–61.
20. Miodownik, *Stuff Matters*, 61.
21. Sara Wermiel, "California Concrete, 1876–1906: Jackson, Percy, and the Beginnings of Reinforced Concrete Construction in the United States," *Proceedings of the Third International Congress on Construction History*, May 2009.
22. Ernest Ransome and Alexis Saurbrey, *Reinforced Concrete Buildings* (New York: McGraw-Hill, 1912), 1.
23. Bay Area Census, http://www.bayareacensus.ca.gov/counties/SanFrancisco County40.htm.
24. Courland, *Concrete Planet*, 3190.
25. "A Boom in the Artificial Stone Trade," *San Francisco Chronicle*, December 24, 1885.
26. Wermiel, "California Concrete," 2–4.
27. Ransome and Saurbrey, *Reinforced Concrete Buildings*, 3.
28. Reyner Banham, *A Concrete Atlantis: U.S. Industrial Building and European Modern Architecture* (Boston: MIT Press, 1989), 2.
29. Ransome and Saurbrey, *Reinforced Concrete Buildings*, 163–64.
30. Wermiel, "California Concrete," 7.
31. "Would Prohibit Concrete Buildings," *Los Angeles Times*, October 23, 1905.
32. *The Brickbuilder* 15, no. 5 (May 1906).
33. Bekins Company History, http://www.fundinguniverse.com/company-histories/bekins-company-history/.
34. Courland, *Concrete Planet*, 4522–524.
35. Ibid., 4433–440.
36. Ibid., 4432–433, 4475, 4504–518, 4547, 4556.
37. Wm. Hom Hall, "Some Lessons of the Earthquake and Fire," *San Francisco Chronicle*, June 1, 1906.
38. "Blow Aimed at Concrete," *Los Angeles Times*, June 13, 1906.
39. "Building May Be Retarded," *San Francisco Chronicle*, March 3, 1907.
40. "The Cement Age," *Healdsburg Tribune*, February 28, 1907.
41. Wermiel, "California Concrete," 7.

NOTES

42. C. C. Carlton, "Edison Tells How a House Can Be 'Cast,'" *San Francisco Call*, December 23, 1906.
43. Courland, *Concrete Planet*, 3447–449.
44. "The Advantages and Limitations of Reinforced Concrete," *Scientific American*, May 12, 1906, 383.
45. Amy E. Slaton, *Reinforced Concrete and the Modernization of American Building, 1900–1930* (Baltimore, MD: Johns Hopkins University Press, 2001), 19.
46. "Conquest of Mixture Soon to Be Complete," *Los Angeles Herald*, November 15, 1908.
47. Tom Lewis, *Divided Highways: Building the Interstate Highways, Transforming American Life* (Ithaca, NY: Cornell University Press, 2013), Kindle Location 1064.
48. "Sand and Gravel (Construction) Statistics," US Geological Survey, http://minerals.usgs.gov/minerals/pubs/historical-statistics/ds140-sandc.pdf.
49. "Nassau County Growth," *New York Times*, June 23, 1912.
50. Sidney Redner, "Distribution of Populations," http://physics.bu.edu/~redner/projects/population/cities/chicago.html.
51. Joan Cook, "Henry Crown, Industrialist, Dies," *New York Times*, August 16, 1990.
52. Edwin A. R. Trout, "The German Committee for Reinforced Concrete, 1907–1945," *Construction History*, 2014. https://www.jstor.org/stable/43856074?seq=1#page_scan_tab_contents.
53. L. W.-C. Lai, K. W. Chau, and F. T. Lorne, "The Rise and Fall of the Sand Monopoly in Colonial Hong Kong," *Ecological Economics* 128 (2016): 106–116.
54. "Hoover Dam Aggregate Classification Plant," *Historic American Engineering Record*, July 2009, 13.
55. "Hoover Dam Aggregate Classification Plant," 8.
56. Courland, *Concrete Planet*, 3511–512.
57. Megan Chusid, "How One Simple Material Shaped Frank Lloyd Wright's Guggenheim," https://www.guggenheim.org/blogs/checklist/how-one-simple-material-shaped-frank-lloyd-wrights-guggenheim.

Chapter 3: Paved with Good Intentions

1. Dwight D. Eisenhower, *At Ease: Stories I Tell to Friends* (Doubleday, 1967), 155.
2. Christopher Klein, "The Epic Road Trip That Inspired the Interstate Highway System," *History*, history.com/news/the-epic-road-trip-that-inspired-the-interstate-highway-system.
3. Eisenhower, *At Ease*, 157.
4. "Highways History, Part 1," *Greatest Engineering Achievements of the 20th Century*, National Academy of Engineering, http://www.greatachievements.org/?id=3790.
5. Henry Petroski, *The Road Taken: The History and Future of America's Infrastructure* (New York: Bloomsbury, 2016), 43.
6. Eisenhower, *At Ease*, 158.

7. Dwight D. Eisenhower, "Eisenhower's Army Convoy Notes 11-3-1919"; https://www.fhwa.dot.gov/infrastructure/convoy.cfm.

8. Earl Swift, *The Big Roads: The Untold Story of the Engineers, Visionaries, and Trailblazers Who Created the American Superhighway* (Boston: Houghton Mifflin Harcourt, 2011), Kindle Location 1006.

9. Eisenhower, *At Ease*, 167.

10. Vaclav Smil, *Making the Modern World: Materials and Dematerialization* (Hoboken, NJ: Wiley, 2013), 54.

11. "Materials in Use in U.S. Interstate Highways," US Geological Survey, October 2006.

12. Tom Lewis, *Divided Highways: Building the Interstate Highways, Transforming American Life* (Ithaca, NY: Cornell University Press, 2013), 2.

13. Rickie Longfellow, "Back in Time: Building Roads," *Highway History*, Federal Highway Administration, https://www.fhwa.dot.gov/infrastructure/back 0506.cfm.

14. Petroski, *The Road Taken*, 3–4.

15. "Learn About Asphalt," BeyondRoads.com, Asphalt Education Partnership, http://www.beyondroads.com/index.cfm?fuseaction=page&filename =history.html.

16. Peter Mikhailenko, "Valorization of By-products and Products from Agro-Industry for the Development of Release and Rejuvenating Agents for Bituminous Materials," unpublished doctoral thesis, Université de Toulouse, 2015, 13.

17. Carole Simm, "The History of the Pitch Lake in Trinidad," *USA Today*, http://traveltips.usatoday.com/history-pitch-lake-trinidad-58120.html.

18. Maxwell Gordon Lay, "Roads and Highways," *Encyclopedia Britannica*, https://www.britannica.com/technology/road.

19. Bill Davenport, Gerald Voigt, and Peter Deem, "Concrete Legacy: The Past, Present, and Future of the American Concrete Pavement Association," American Concrete Pavement Association, 2014, 11.

20. "How flat can a highway be?" Portland Cement Association, 1959.

21. "The United States has about 2.2 million miles of paved roads . . ." Asphalt Pavement Alliance, http://www.asphaltroads.org/why-asphalt/economics/.

22. "World Asphalt (Bitumen)," Freedonia Group, November 2015.

23. Swift, *The Big Roads*, 457.

24. Lewis, *Divided Highways*, 719–21.

25. "Highways," Portland Cement Association, http://www.cement.org/concrete -basics/paving/concrete-paving-types/highways.

26. Swift, *The Big Roads*, 197–203.

27. Ibid., 247–53.

28. Lewis, *Divided Highways*, 1042–44.

29. Davenport, et al., "Concrete Legacy," 13.

30. Lewis, *Divided Highways*, 339–49, 532.

31. "Land Reclamation and Highway Development Must Go Together," *Water & Sewage Works*, Vol. 55 (Scranton Publishing Company, 1918).

32. J. D. Pierce, "Sand and Gravel in Illinois," *The National Sand and Gravel Bulletin*, 1921, 29.

33. Davenport, et al., "Concrete Legacy," 17.

34. "Roads," Encyclopedia.com, http://www.encyclopedia.com/topic/Roads.aspx

35. Lewis, *Divided Highways*, 971–73.

36. Kurt Snibbe, "Back in the Day: Road Camp Prisoners Built Roads," *The Press-Enterprise*, January 18, 2013, and "History of the North Carolina Correction System," North Carolina Department of Public Safety, http://www.doc.state.nc.us/admin/page1.htm.

37. Mark S. Foster, *Henry J. Kaiser: Builder in the Modern American West* (Austin: University of Texas Press, 2012), 5, 7.

38. Wes Starratt, "Sand Castles," *San Francisco Bay Crossings*, June 2002; http://www.baycrossings.com/dispnews.php?id=1083.

39. Foster, *Henry J. Kaiser*, 10.

40. Albert P. Heiner, *Henry J. Kaiser: Western Colossus* (Halo Books, 1991), 6–7.

41. "Six Million Dollar Arroyo Parkway Opened," *Los Angeles Times*, December 31, 1940; and "A Look at the History of the Federal Highway Administration," Federal Highway Administration, https://www.fhwa.dot.gov/byday/fhbd1230.htm.

42. David Irving, *Hitler's War* (London: Focal Point Publications, 2001), 769; http://www.jrbooksonline.com/PDF_Books_added2009-2/HW1.pdf.

43. Eisenhower, *At Ease*, 166–7.

44. Lewis and Swift each discuss the history of the campaign for a national highway system in considerable depth.

45. Richard F. Weingroff, "The Year of the Interstate," *Public Roads*, January–February 2006.

46. "The Size of the Job," *Highway History*, Federal Highway Administration, https://www.fhwa.dot.gov/infrastructure/50size.cfm.

47. Wallace W. Key, Annie Laurie Mattila, "Sand and Gravel," *Minerals Yearbook 1958*, US Bureau of Mines.

48. Author interviews and "Rogers Group at 100," *Aggregates Manager*, November 1, 2008.

49. Swift, *The Big Roads*, 3002.

50. Lewis, *Divided Highways*, 2532.

51. Swift, *The Big Roads*, 3663.

52. "The Interstate Highway System—Facts & Summary," *History.com*, http://www.history.com/topics/interstate-highway-system.

53. Weingroff, "The Year of the Interstate."

54. "Interstate Frequently Asked Questions," Federal Highway Administration, http://www.fhwa.dot.gov/interstate/faq.cfm.

55. Ibid., and Swift, *The Big Roads*, 3848.

56. "The United Nations and Road Safety," United Nations, http://www.un.org/en/roadsafety/.

57. Lewis, *Divided Highways*, 115–120.

58. "Our Nation's Highways 2011," Federal Highway Administration, 25.

59. "Roads," Encyclopedia.com, http://www.encyclopedia.com/topic/Roads.aspx.

60. "Our Nation's Highways," 36.

61. Ibid., 44.

62. Mark S. Kuhar and Josephine Smith, "Rock Through the Ages: 1896–2016," *Rock Products*, July 13, 2016; http://www.rockproducts.com/features/15590 -rock-through-the-ages-1896-2016.html#.WAL4kJMrLdQ.

63. "U.S. Swimming Pool and Hot Tub Market 2015," Association of Pool and Spa Professionals.

64. "Sand and Gravel (Construction) Statistics," US Geological Survey, April 1, 2014.

65. "Rock Products 120th Anniversary," *Rock Products*, December 22, 2015; www.rockproducts.com/blog/120th-anniversary/14999-rock-products -120th-anniversary-part-6.html.

66. "Our Nation's Highways," 4.

67. "Traffic Gridlock Sets New Records," Texas A&M University press release, August 26, 2015.

68. "Global Land Transport Infrastructure Systems," International Energy Agency, 2013, 12.

69. Ibid., 6.

Chapter 4: The Thing That Lets Us See Everything

1. John Douglas, "Glass Sand Mining," *e-WV: The West Virginia Encyclopedia*, August 7, 2012.

2. Quentin Skrabec Jr., *Michael Owens and the Glass Industry* (Gretna, LA: Pelican, 2006), 66.

3. Ibid., 76–78.

4. Barbara L. Floyd, *The Glass City: Toledo and the Industry That Built It* (Ann Arbor: University of Michigan Press, 2014), 49–50.

5. For a detailed explanation on the manufacture of different types of glass, see Alan Macfarlane and Gerry Martin, *The Glass Bathyscaphe: How Glass Changed the World* (Profile Books, 2011), Appendix 1.

6. Mark Miodownik, *Stuff Matters: Exploring the Marvelous Materials That Shape Our Man-Made World* (Boston: Houghton Mifflin Harcourt, 2014), 141.

7. Macfarlane and Martin, *The Glass Bathyscaphe*, Kindle Locations 148-156.

8. Skrabec, *Owens*, 21.

9. Miodownik, *Stuff Matters*, 144–147.

10. Michael Welland, *Sand: The Never-Ending Story* (Berkeley: University of California Press, 2009), 248.

11. Vincent Ilardi, *Renaissance Vision from Spectacles to Telescopes*. Memoirs of the American Philosophical Society, v. 259 (Philadelphia: American Philosophical Society, 2007), 182.

12. Macfarlane and Martin, *The Glass Bathyscaphe*, 1747–752.

13. Richard Dunn, *The Telescope: A Short History*, reprint ed. (New York: Conway, 2011), 22.

14. Ilardi, *Renaissance Vision*, 182.

15. Laura J. Snyder, *Eye of the Beholder: Johannes Vermeer, Antoni van Leeuwenhoek, and the Reinvention of Seeing* (New York: W. W. Norton, 2016), 6.

16. Snyder, *Eye of the Beholder*, 104.

17. Welland, *Sand*, 16–17.

18. Snyder, *Eye of the Beholder*, 4.
19. Skrabec, *Michael Owens*, 49.
20. Welland, *Sand*, 248.
21. Floyd, *The Glass City*, 18–19.
22. Ibid., 1.
23. Skrabec, *Michael Owens*, 124.
24. Floyd, *The Glass City*, 28–29.
25. Skrabec, *Michael Owens*, 14–15.
26. Ibid., 14–15 and 88–89.
27. "The American Society of Mechanical Engineers Designates the Owens 'AR' Bottle Machine as an International Historic Engineering Landmark," *American Society of Mechanical Engineers*, May 17, 1983; https://www.asme.org/getmedia/a9e54878-05b1-4a91-a027-fe3b7e08699e/86-Owens-AR-Bottle-Machine.aspx.
28. Floyd, *The Glass City*, 48.
29. "Sand and Gravel (Industrial) Statistics," US Geological Survey, 2016.
30. Kenneth Schoon, "Sand Mining in and around Indiana Dunes National Lake Shore," National Parks Service, May 2015. https://www.nps.gov/rlc/great-lakes/sand-mining-in-indiana-dunes.htm.
31. "Vanishing Lake Michigan Sand Dunes: Threats from Mining," Lake Michigan Federation, date unknown.
32. Schoon, "Sand Mining."
33. "The Largest Glass Sand Plant in the Country," *Rock Products and Building Materials*, April 7, 1914, 36.
34. Skrabec, *Michael Owens*, 80.
35. "History of Bottling," Coca-Cola Company, http://www.coca-colacompany.com/our-company/history-of-bottling.
36. Floyd, *The Glass City*, 105.
37. Vaclav Smil, *Making the Modern World: Materials and Dematerialization* (Hoboken, NJ: Wiley, 2013), 92.
38. "World Flat Glass Market Report," Freedonia Group, August 2016.
39. "About O-I," Owens-Illinois, http://www.o-i.com/About-O-I/Company-Facts/.
40. "World Flat Glass Market Report," Freedonia Group, August 2016.

Chapter 5: High Purity, High Tech

1. Alex Glover, "A Brief Review of the History, Geology and Modern Uses of the Minerals Mined in the Spruce Pine Mining District." http://owacc.com/land%20sales/345%20AVERY/345.pdf.
2. David Biddix and Chris Hollifield, *Images of America: Spruce Pine* (Mt. Pleasant, SC: Arcadia Publishing, 2009), 9.
3. Ibid., 10.
4. John W. Schlanz, "High Pure and Ultra High Pure Quartz," *Industrial Minerals and Rocks*, 7th ed. (Society for Mining, Metallurgy, and Exploration, March 5, 2006), 833–37.
5. Harris Prevost, "Spruce Pine Sand and the Nation's Best Bunkers," *North Carolina's High Country Magazine*, July 2012.

6. David O. Woodbury, *The Glass Giant of Palomar* (New York: Dodd, Mead, 1970), 185.

7. Joel Shurkin, *Broken Genius: The Rise and Fall of William Shockley, Creator of the Electronic Age* (New York: Macmillan Science, 2006), 171.

8. Vaclav Smil, *Making the Modern World: Materials and Dematerialization* (Hoboken, NJ: Wiley, 2013), 40.

9. For this summary of the extremely complex process of creating silicon, two excellent sources were Eric Williams's "Global Production Chains and Sustainability: The case of high-purity silicon and its applications in IT and renewable energy," a report published in 2000 by the United Nations University /Institute of Advanced Studies; and the Quartz Corporation's website, including "Polysilicon Production," http://www.thequartzcorp.com/en/blog/2014 /04/28/polysilicon-production/61.

10. "Silicon," *Mineral Industry Surveys*, December 2016, US Geological Survey, March 2017.

11. "Polysilicon pricing and the Chinese market," Quartz Corp, June 14, 1016; http://www.thequartzcorp.com/en/blog/2016/06/14/polysilicon-pricing-and -the-chinese-solar-market/186.

12. "Crucibles," Quartz Corp, http://www.thequartzcorp.com/en/applications /crucibles.html.

13. Jessica Roberts, "High purity quartz: under the spotlight," *Industrial Minerals*, December 1, 2011.

14. Schlanz, "High Pure and Ultra High Pure Quartz," 1–2.

15. Reiner Haus, Sebastian Prinz, and Christoph Priess, "Assessment of High Purity Quartz Resources," *Quartz: Deposits, Mineralogy and Analytics* (Springer Geology, 2012), chapter 2.

16. Prevost, "Spruce Pine Sand and the Nation's Best Bunkers."

17. Affidavit of Thomas Gallo, PhD, *Unimin Corporation v. Thomas Gallo and I-Minerals USA*, Mitchell County Superior Court, North Carolina, July 12, 2014.

18. "High purity quartz: a cut above," *Industrial Minerals*, December 2013, 22.

19. "High Purity Quartz Crucibles: Part I," Quartz Corp, November 28, 2016; http://www.thequartzcorp.com/en/blog/2016/11/28/high-purity-quartz -crucibles-part-i/218.

20. "How Microchips Are Made," Science Channel, https://www.youtube.com /watch?v=F2KcZGwntgg.

21. Smil, *Making the Modern World*, 74.

22. "From Sand to Circuits: How Intel Makes Chips," *Intel*, date unknown.

23. "Semiconductor Manufacturing Process," Quartz Corp, January 13, 2014; http://www.thequartzcorp.com/en/blog/2014/01/13/semiconductor -manufacturing-process/42.

24. Konstantinos I. Vatalis, George Charalambides, and Nikolas Ploutarch Benetis, "Market of High Purity Quartz Innovative Applications," *Procedia Economics and Finance* 24 (2015): 734–42. Part of special issue: International Conference on Applied Economics, July 2–4, 2015, Kazan, Russia.

25. Affidavit of Richard Zielke, *Unimin Corporation v. Thomas Gallo and I-Minerals USA*, Mitchell County Superior Court, North Carolina, July 25, 2014.

26. "Quick Facts: Mitchell County, North Carolina," US Census Bureau, http://www.census.gov/quickfacts/table/PST045215/37121.

27. Rich Miller, "The Billion Dollar Data Centers," Data Center Knowledge, April 29, 2013; http://www.datacenterknowledge.com/archives/2013/04/29/the-billion-dollar-data-centers/.

Chapter 6: Fracking Facilitator

1. Leonardo Maugeri, "Oil: The Next Revolution," Harvard Kennedy School /Belfer Center for Science and International Affairs, June 2012, 53.

2. "How much shale gas is produced in the United States?" US Energy Information Administration; https://www.eia.gov/tools/faqs/faq.php?id=907&t=8.

3. Maugeri, "Oil," 56–57.

4. Don Bleiwas, "Estimates of Hydraulic Fracturing (Frac) Sand Production, Consumption, and Reserves in the United States," *Rock Products* 118, no. 5 (May 2015).

5. "Silica Sand Mining in Wisconsin," Wisconsin Department of Natural Resources, January 2012, 4–5.

6. Stephanie Porter, "Breaking the Rules for Profit," Land Stewardship Project, November 26, 2014, 4.

7. Bleiwas, "Estimates of Hydraulic . . ."

8. "Sand and Gravel (Industrial)," *US Geological Survey Mineral Commodity Summaries*, January 2017, 144.

9. Thomas P. Dolley, "Silica," *US Geological Survey 2014 Minerals Yearbook*, 66.1.

10. "Silica Sand Mining in Wisconsin," Wisconsin Department of Natural Resources, January 2012, 8.

11. "High Capacity Wells," Wisconsin Department of Natural Resources; http://dnr.wi.gov/topic/Wells/HighCap/.

12. Steven Verburg, "Frac sand miners fined $60,000 for stormwater spill in creek," Madison.com, September 9, 2014; http://host.madison.com/news/local/environment/frac-sand-miners-fined-for-stormwater-spill-in-creek/article_49ceb1e1-87eb-5177-887d-4d03b75b4c88.html.

13. Emily Chapman, et al., "Communities at Risk: Frac Sand Mining in the Upper Midwest," *Boston Action Research*, September 25, 2014.

14. Ali Mokdad, et al., "Actual Causes of Death in the United States, 2000," *JAMA* 291, no. 10 (March 10, 2004): 1238–45.

15. E. J. Esswein, et al., "Occupational exposures to respirable crystalline silica during hydraulic fracturing," *Journal of Occupational and Environmental Hygiene* 10, no. 7 (2013): 347–56; https://www.ncbi.nlm.nih.gov/pubmed/23679563.

16. Soren Rundquist and Bill Walker, "Danger in the Air," Environmental Working Group, September 25, 2014; http://www.ewg.org/research/danger-in-the-air#.WekhBBOPLdQ.

17. Soren Rundquist, "Danger in the Air," Part 2, Environmental Working Group, September 25, 2014; http://www.ewg.org/research/sandstorm/health-concerns-silica-outdoor-air#.WekhMhOPLdQ.

18. John Richards and Todd Brozell, "Assessment of Community Exposure to Ambient Respirable Crystalline Silica near Frac Sand Processing Facilities," *Atmosphere* 6 *(*July 24, 2015): 960–82.
19. Chapman, "Communities at Risk," 10–11.
20. Porter, "Breaking the Rules," 4.
21. Ibid., 15.
22. Ibid., 6.
23. Steven Verburg, "Scott Walker, Legislature altering Wisconsin's way of protecting natural resources," Madison.com, October 4, 2015.
24. "Silica Sand Mines in Minnesota," Minnesota Department of Natural Resources, 2016.
25. Karen Zamora and Josephine Marcotty, "Winona County passes frac sand ban, first in the state to take such a stand," *Star Tribune,* November 22, 2016; http://www.startribune.com/winona-county-passes-frac-sand-ban-first-in-the-state-to-take-such-a-stand/402569295/.
26. Thomas W. Pearson, *When the Hills Are Gone: Frac Sand Mining and the Struggle for Community* (Minneapolis: University of Minnesota Press, 2017), 4.
27. Leighton Walter Kille, "The environmental costs and benefits of fracking: The state of research," *Journalist's Resource;* http://journalistsresource.org/studies/environment/energy/environmental-costs-benefits-fracking.
28. "Global Trends 2030: Alternative Worlds," National Intelligence Council, December 2012, 57.

Chapter 7: Miami Beach-Less

1. Ryan McNeill, Deborah J. Nelson, and Duff Wilson, "Water's edge: the crisis of rising sea levels," Reuters, September 4, 2014; https://www.reuters.com/investigates/special-report/waters-edge-the-crisis-of-rising-sea-levels/.
2. "Disappearing Beaches: Modeling Shoreline Change in Southern California," US Geological Survey, March 27, 2017.
3. Orrin H. Pilkey Jr. and J. Andrew G. Cooper, *The Last Beach* (Durham, NC: Duke University Press, 2014), 14.
4. Michael Welland, *Sand: The Never-Ending Story* (Berkeley: University of California Press, 2009), 18.
5. Patrick Reilly, "Without more sand, SoCal stands to lose big chunk of its beaches," *Christian Science Monitor,* March 28, 2017.
6. Bob Marshall, "Losing Ground: Southeast Louisiana Is Disappearing, Quickly," *Scientific American,* August 28, 2014.
7. Edward J. Anthony, et al., "Linking rapid erosion of the Mekong River delta to human activities," *Nature.com Scientific Reports* 5, article no. 14745, October 8, 2015.
8. Pilkey and Cooper, *The Last Beach,* 25–28, 30, 32–33.
9. Pedro A. Gelabert, "Environmental Effects of Sand Extraction Practices in Puerto Rico," papers presented at a UNESCO–University of Puerto Rico workshop entitled "Integrated Framework for the Management of

Beach Resources within the Smaller Caribbean Islands," October 21–25, 1996.

10. Pilkey and Cooper, *The Last Beach*, 37–38.

11. Desmond Brown, "Facing Tough Times, Barbuda Continues Sand Mining Despite Warnings," Inter Press Service News Agency, June 22, 2013.

12. Email correspondence with Dr. Amy E. Potter, Assistant Professor of Geography, Department of History, Armstrong State University.

13. Jase D. Ousley, Elizabeth Kromhout, and Matthew H. Schrader, "Southeast Florida Sediment Assessment and Needs Determination (SAND) Study," US Army Corps of Engineers, August 2013, 93.

14. Lisa Broad, "Treasure Coast fighting Miami-Dade efforts to ship its sand south," *Stuart News/Port St. Lucie News*, September 20, 2015.

15. John Branch, "Copacabana's Natural Sand Is Just Right for Olympic Beach Volleyball," *New York Times*, August 9, 2016.

16. Pilkey and Cooper, *The Last Beach*, xi.

17. John R. Gillis, *The Human Shore: Seacoasts in History*, reprint ed. (Chicago: University of Chicago Press, 2015), 155.

18. Tatyana Ressetar, "The Seaside Resort Towns of Cape May and Atlantic City, New Jersey Development, Class Consciousness, and the Culture of Leisure in the Mid to Late Victorian Era," thesis, University of Central Florida, 2011; http://stars.library.ucf.edu/etd/1704/.

19. Ibid., 16.

20. D. J. Waldie, "How Angelenos invented the L.A. summer—in the beginning was the barbecue," *Los Angeles Times*, July 9, 2017.

21. Gillis, *The Human Shore*, 160–61.

22. T. D. Allman, *Finding Florida: The True History of the Sunshine State* (New York: Grove Press, 2014), 319–20, 333.

23. "History of Broward County," http://www.broward.org/History/Pages/BCHistory.aspx.

24. Allman, *Finding Florida*, 337.

25. David Fleshler, "Wade-ins ended beach segregation," *Sun Sentinel*, April 13, 2015.

26. "Important Broward County Milestones," http://www.broward.org/History/Pages/Milestones.aspx.

27. Allman, *Finding Florida*, 347.

28. Robert L. Wiegel, "Waikiki Beach, Oahu, Hawaii: History of its transformation from a natural to an urban shore," *Shore & Beach*, Spring 2008.

29. Pilkey and Cooper, *The Last Beach*, 168.

30. James McAuley, "Fake Seine beaches are part of a Paris summer. This year, they're making officials nervous," *Washington Post*, July 28, 2016.

31. René Kolman, "New Land by the Sea: Economically and Socially, Land Reclamation Pays," International Association of Dredging Companies, May 2012; https://www.iadc-dredging.com/ul/cms/fck-uploaded/documents/PDF%20Articles/article-new-land-by-the-sea.pdf.

32. "Fijian Economy," Fiji High Commission to the United Kingdom, http://www.fijihighcommission.org.uk/about_3.html.

33. McNeill, et al., "Water's edge: the crisis of rising sea levels."
34. Justin Gillis, "Flooding of Coast, Caused by Global Warming, Has Already Begun," *New York Times*, September 3, 2016.
35. Gillis, *The Human Shore*, 12, 184.
36. Dylan E. McNamara, Sathya Gopalakrishnan, Martin D. Smith, and A. Brad Murray, "Climate Adaptation and Policy-Induced Inflation of Coastal Property Value," *PLoS One* 10, no. 3 (March 25, 2015).
37. McNeill, et al., "Water's edge: the crisis of rising sea levels."
38. JoAnne Castagna, "Messages in the sand from Hurricane Sandy," US Army Corps of Engineers, September 7, 2016; https://www.dvidshub.net/news/208990/messages-sand-hurricane-sandy.
39. Pilkey and Cooper, *The Last Beach*, 70.
40. Ousley, et al., "Southeast Florida Sediment Assessment and Needs Determination (SAND) Study," 93.
41. "Beach Nourishment Viewer," Program for the Study of the Developed Shoreline, Western Carolina University; http://beachnourishment.wcu.edu/.
42. Welland, *Sand*, 123.
43. Pilkey and Cooper, *The Last Beach*, 16–18, 21, 83–85.
44. Andres David Lopez, "Study: Sand nourishment linked to fewer marine life," *Palm Beach Daily News*, April 4, 2016.
45. Sammy Fretwell, "Marine life dwindles after beach renourishment at Folly, report says," *The State*, August 19, 2016.
46. Steve Lopez, "A dangerous confluence on the California coast: beach erosion and sea level rise," *Los Angeles Times*, August 24, 2016.

Chapter 8: Man-Made Lands

1. Email correspondence with René Kolman, secretary general of the International Association of Dredging Companies, March 21, 2017.
2. A.G.M.Groothuizen, "World Development and the Importance of Dredging," *PIANC Magazine*, January 2008.
3. "Making Up Ground," 99% Invisible, September 15, 2016. http://99percentinvisible.org/episode/making-up-ground.
4. "Chicago Shoreline History," City of Chicago, http://www.cityofchicago.org/dam/city/depts/cdot/ShorelineHistory.pdf, date unknown.
5. Brent Ryan et al., "Developing the Littoral Gradient," MIT Center for Advanced Urbanism, Fall 2015.
6. René Kolman, "New Land by the Sea: Economically and Socially, Land Reclamation Pays," International Association of Dredging Companies, May 2012.
7. Kolman, "New Land by the Sea."
8. "Beyond Sand and Sea," International Association of Dredging Companies, 2015.
9. Ryan, "Developing the Littoral Gradient."
10. "Shifting Sand: How Singapore's demand for Cambodian sand threatens ecosystems and undermines good governance," *Global Witness*, May 2010.
11. Samanth Subramanian, "How Singapore Is Creating More Land for Itself," *New York Times*, April 20, 2017.

12. Alister Doyle, "Coastal land expands as construction outpaces sea level rise," *Reuters*, August 25," 2016.

13. "Beyond Sand and Sea," 50.

14. For this short history of Dubai, I relied heavily on Jim Krane's *City of Gold: Dubai and the Dream of Capitalism* (New York: St. Martin's Press, 2009).

15. Ibid., 4.

16. Ibid., 28–29.

17. Ibid., 70.

18. Gargi Kapadia, "Palm Island Construction with Management 5 Ms," Welingkar Institute of Management Development and Research, August 12, 2013.

19. "Palm Islands, Dubai—Compression of the Soil," *CDM Smith*, date unknown.

20. Krane, *City of Gold*, 154.

21. "Palm Islands, Dubai—Compression of the Soil."

22. Adam Luck, "How Dubai's $14 billion dream to build The World is falling apart," *Daily Mail*, April 11, 2010.

23. Tida Choomchaiyo, "The Impact of the Palm Islands," https://sites.google.com/site/palmislandsimpact/environmental-impacts/long-term, December 5, 2009.

24. Krane, *City of Gold*, 230.

25. David Medio, "Persian Gulf: The Cost of Coastal Development to Reefs," World Resources Institute, http://www.wri.org/persian-gulf-cost-coastal-development-reefs.

26. John A. Burt, "The environmental costs of coastal urbanization in the Arabian Gulf," *City: analysis of urban trends, culture, theory, policy, action* 18, no. 6 (November 28, 2014): 760–770.

27. Wei Wang, Hui Liu, Yongqi Li, and Jilan Su. "Development and Management of Land Reclamation in China," *Ocean & Coastal Management*, Volume 102, Part B, December 2014, 415–25.

28. Krane, *City of Gold*, 224.

29. "Asia-Pacific Maritime Security Strategy," US Department of Defense, August 2015, 9.

30. Ibid., 19.

31. Tian Jun-feng, et al., "Review of the ten-year development of Chinese Dredging Industry," *Port and Waterway Engineering,* January 2013.

32. Andrew S. Erickson and Kevin Boyd, "Dredging Under the Radar: China Expands South Sea Foothold," *The National Interest,* August 26, 2015, and Carrie Gracie, "What is China's 'magic island-making' ship?," BBC, November 6, 2017. bbc.com/news/world-asia-china-41882081.

33. "In the Matter of the South China Sea Arbitration," Permanent Court of Arbitration, July 12, 2016, 352.

34. "Asia-Pacific Maritime Security Strategy," 19–21.

35. "In the Matter of the South China Sea Arbitration," 416.

36. Greg Torode, "'Paving paradise': Scientists alarmed over China island building in disputed sea," *Reuters*, June 25, 2015.

37. Agence France-Presse, "China's plans to expand in the South China Sea with a floating nuclear power plant continue," *Mercury*, December 25, 2017. http://

www.themercury.com.au/technology/chinas-plans-to-expand-in-the-south-china-sea-with-a-floating-nuclear-power-plant-continue/news-story/bdc1bf6f6b556daf097b3199b5690182.

38. David E. Sanger, "Piling Sand in a Disputed Sea, China Literally Gains Ground," *New York Times*, April 9, 2015.

39. Hrvoje Hranjski and Jim Gomez, "China rejects freeze on island building; ASEAN divided," Associated Press, August 16, 2015.

40. David Brunnstrom and Matt Spetalnick, "Tillerson says China should be barred from South China Sea islands," Reuters, January 12, 2017.

41. Benjamin Haas, "Steve Bannon: 'We're going to war in the South China Sea . . . no doubt,'" *Guardian*, February 1, 2017.

42. Mike Morgan, *Sting of the Scorpion: The Inside Story of the Long Range Desert Group* (Stroud, Glouchestershire: The History Press, 2011), Kindle Locations 401, 500.

43. Trevor Constable, "Bagnold's Bluff: The Little-Known Figure Behind Britain's Daring Long Range Desert Patrols," *The Journal of Historical Review* 18, no. 2 (March/April 1999).

Chapter 9: Desert War

1. "SCIO news briefing on the 5th national monitoring survey of desertification and sandification," State Council Information Office press release, December 31, 2015.

2. W. Chad Futrell, "A Vast Chinese Grassland, a Way of Life Turns to Dust," *Circle of Blue*, January 21, 2008.

3. "An Introduction to the United Nations Convention to Combat Desertification," http://www.unccd.int/Lists/SiteDocumentLibrary/Publications/factsheets-eng.pdf.

4. Fred Attewill, "Stopping the Sands of Time," *Metro* (UK), January 18, 2012; http://metro.co.uk/2012/01/18/stopping-the-sands-of-time-plans-to-stem-the-tide-of-advancing-deserts-289361/.

5. "SCIO news briefing."

6. Hong Jiang, "Taking Down the Great Green Wall: The Science and Policy Discourse of Desertification and Its Control in China," in *The End of Desertification? Disputing Environmental Change in the Drylands*, Roy Behnke and Michael Mortimore, eds. (Springer, 2016), 513–36.

7. Diana K. Davis, *The Arid Lands: History, Power, Knowledge* (Cambridge, MA: MIT Press, 2016), 7.

8. Elion Resources Group, "Elion's Ecosystem," 2013.

9. X. M. Wang, et al., "Has the Three Norths Forest Shelterbelt Program solved the desertification and dust storm problems in arid and semiarid China?" *Journal of Arid Environments* 74, no. 1 (January 2010): 13–22.

10. Jiang, "Taking Down the Great Green Wall."

11. Shixiong Cao, et al., "Damage Caused to the Environment by Reforestation Policies in Arid and Semi-Arid Areas of China," *AMBIO: A Journal of the Human Environment* 39, no. 4 (June 2010): 279–83.

12. Weimin Xi, et al., "Challenges to Sustainable Development in China: A Review of Six Large-Scale Forest Restoration and Land Conservation Programs," *Journal of Sustainable Forestry* 33 (2014): 435–53.
13. Wang, et al., "Has the Three Norths . . ."

Chapter 10: Concrete Conquers the World

1. Taras Grescoe, "Shanghai Dwellings Vanish, and With Them, a Way of Life," *New York Times*, January 23, 2017.
2. "Basic Statistics on National Population Census," Shanghai Municipal Bureau of Statistics, http://www.stats-sh.gov.cn/tjnj/nje11.htm?d1=2011tjnje/E0226 .htm.
3. John E. Fernández, "Resource Consumption of New Urban Construction in China," *Journal of Industrial Ecology* 11, no. 2 (April 2007): 99–115.
4. Chen Xiqing, et al., "In-channel sand extraction from the mid-lower Yangtze channels and its management: Problems and challenges," *Journal of Environmental Planning and Management* 49, no. 2 (2006): 309–20.
5. Xijun Lai, David Shankman, et al., "Sand mining and increasing Poyang Lake's discharge ability: A reassessment of causes for lake decline in China," *Journal of Hydrology* 519 (2014): 1698–706.
6. Concrete Sustainability Council, http://www.concretesustainabilitycouncil .org/index.php?pagina=rss/pagina1.
7. Robert Courland, *Concrete Planet: The Strange and Fascinating Story of the World's Most Common Man-Made Material* (Amherst, NY: Prometheus Books, 2011), Kindle Locations 183–86.
8. "Sustainable Cities and Communities," United Nations Development Programme, http://www.undp.org/content/undp/en/home/sdgoverview/post-2015 -development-agenda/goal-11.html.
9. "Global Trends: Alternative Worlds," US National Intelligence Council, December 2012, 9.
10. Fernandez, "Resource Consumption of New Urban Construction in China," 2.
11. Courland, *Concrete Planet*, 3914–916.
12. Fernandez, "Resource Consumption of New Urban Construction in China," 5–7.
13. Charles Kenny, "Paving Paradise," *Foreign Policy*, January 3, 2012.
14. Alex Barnum, "First-of-Its-Kind Index Quantifies Urban Heat Islands," California Environmental Protection Agency press release, September 16, 2015.
15. Courland, *Concrete Planet*, 4758–4763.
16. Ian Boost, "Houston's Flood Is a Design Problem," *TheAtlantic.com*, August 28, 2017. https://www.theatlantic.com/technology/archive/2017/08/why-cities -flood/538251/.
17. Ryan McNeill, Deborah J. Nelson, and Duff Wilson, "Water's Edge," Reuters, September 4, 2014.
18. "Typical Systems of Reinforced Concrete Construction," *Scientific American*, May 12, 1906, 386.

19. "The Age of Concrete," *San Francisco Chronicle,* January 14, 1906.
20. Ernest Ransome and Alexis Saurbrey, *Reinforced Concrete Buildings* (New York: McGraw-Hill, 1912), 208.
21. Vaclav Smil, *Making the Modern World: Materials and Dematerialization* (Hoboken, NJ: Wiley, 2013), 56.
22. "Special NRC Oversight at Seabrook Nuclear Power Plant: Concrete Degradation," US Nuclear Regulatory Commission, August 4, 2016. http://www.nrc.gov/reactors/operating/ops-experience/concrete-degradation.html.
23. Courland, *Concrete Planet,* 4623–4624.
24. Stephen Farrell, "Iraq: The Wrong Type of Sand," *atwar.blogs.nytimes,* March 31, 2010.
25. "2017 Infrastructure Report Card," American Society of Civil Engineers, 2017, 78.
26. Kevin Sieff, "After billions in U.S. investment, Afghan roads are falling apart," *Washington Post,* January 30, 2014.
27. "2017 Infrastructure Report Card," 17.
28. Ron Nixon, "Human Cost Rises as Old Bridges, Dams and Roads Go Unrepaired," *New York Times,* November 5, 2015.
29. Paul Murphy, "Contextualising China's cement splurge," *FT Alphaville,* October 22, 2014; http://ftalphaville.tumblr.com/post/100653486301/contextualising-chinas-cement-splurge.
30. Smil, *Making the Modern World,* 56.
31. Courland, *Concrete Planet,* 23.

Chapter 11: Beyond Sand

1. The tallies of injuries, displacements, and deaths caused by sand mining, and the specific anecdotes I cite, were compiled by me from English-language local media reports in more than sixty countries. There are undoubtedly many more.
2. Joseph Green, "World demand for construction aggregates to reach 51.7 billion tons," *World Cement,* March 18, 2106.
3. G. Mathias Kondolf, "Hungry Water: Effects of Dams and Gravel Mining on River Channels," *Environmental Management* 21, no. 4 (July 1997): 533–551.
4. C. Howard Nye, "Statement on Behalf of the National Stone, Sand, and Gravel Association before the House Committee on Natural Resources Subcommittee on Energy and Mineral Resources," March 21, 2017.
5. Orrin H. Pilkey Jr. and J. Andrew G. Cooper, *The Last Beach* (Durham, NC: Duke University Press, 2014), 15.
6. "Investigate illegal sand mining in BC," *La Jornada,* December 5, 2015.
7. John G. Parrish, "Aggregate Sustainability in California," *California Geological Survey,* 2012.
8. "Producer Price Index Industry Data: Construction sand and gravel, 1965–2016," US Bureau of Labor Statistics.
9. "World Sand Demand by Region," World Construction Aggregates 2016, Freedonia Group.

10. "Stone, sand, and gravel," *United Nations Comtrade,* https://comtrade
.un.org/.

11. Seol Song Ah, "NK exports 100 tons of sand, gravel, and coal daily from
Sinuiju Harbor," *DailyNK.com,* November 15, 2016.

12. Maxwell Porter, "Beach Sand Mining in St. Vincent and the Grenadines,"
papers presented at a UNESCO–University of Puerto Rico workshop entitled
"Integrated Framework for the Management of Beach Resources within the
Smaller Caribbean Islands," October 21–25, 1996, 142.

13. "Corruption and laundering warrant against two Lafarge officials," ElKhabar
.com, July 7, 2010.

14. "Lafarge Syria alleged to have paid armed groups up to US$100,000/month to
keep cement plant running," *Global Cement,* June 29, 2016, and Alice Bagh-
dijan, "LafargeHolcim CEO's Resignation on Syria Creates Power Vacuum,"
Bloomberg.com, April 23 2017.

15. Global Witness, "Shifting Sand," 2, 7.

16. Sandy Indra Pratama and Denny Armandhanu, "Chep Hernawan: I am also
Candidate to Depart to ISIS," *cnnindonesia.com,* March 19, 2015.

17. Rollo Romig, "How to Steal a River," *New York Times Magazine,* March
1, 2017.

18. Mark Miodownik, *Stuff Matters: Exploring the Marvelous Materials That
Shape Our Man-Made World* (Boston: Houghton Mifflin Harcourt, 2014),
67–70.

19. "Questions and Answers," *BeyondRoads.com,* The Asphalt Education
Partnership; http://www.beyondroads.com/index.cfm?fuseaction=page&file
name=asphaltQandA.html.

20. "Marine Aggregate Extraction: The Need to Dredge: Fact or Fiction?" *Mari-
net,* September 2015; http://www.marinet.org.uk/wp-content/uploads/Marine
-Aggregate-Extraction-The-Need-to-Dredge-Fact-or-Fiction.pdf.

21. "The Phosphorus Challenge," *Phosphorus Futures.* http://phosphorusfutures
.net/the-phosphorus-challenge/.

22. David S. Abraham, *The Elements of Power: Gadgets, Guns, and the Struggle
for a Sustainable Future in the Rare Metal Age* (New Haven: Yale University
Press, 2015), 12.

23. United Nations Environment Programme, "Global Material Flows and Resource Productivity. An Assessment Study of the UNEP International Resource
Panel," 2016.

24. "Living Planet Report 2016," World Wildlife Fund. http://wwf.panda.org
/about_our_earth/all_publications/lpr_2016/.

25. Jim Krane, *City of Gold: Dubai and the Dream of Capitalism* (New York: St.
Martin's Press, 2009). 223–24.

26. *Sandgrains: A Crowdfunded Documentary.* http://sandgrains.org/.

27. For instance, see Bernice Lee, et al., "Resources Futures," *Chatham House,*
December 2012, 2–3, 12, 15.

28. Mark J. Perry, "Today's new homes are 1,000 square feet larger than in 1973,
and the living space per person has doubled over last 40 years," American
Enterprise Institute, February 26, 2014. https://www.aei.org/publication

/todays-new-homes-are-1000-square-feet-larger-than-in-1973-and-the-living
-space-per-person-has-doubled-over-last-40-years/.

29. Richard Dobbs, James Manyika, and Jonathan Woetzel, *No Ordinary Disruption* (New York: Public Affairs, 2016), 8, 94.

30. "Affordable housing key for development and social equality, UN says on World Habitat Day," United Nations press release, October 2, 2017; http://www.un.org/apps/news/story.asp?NewsID=57786#.We_M-ROPLdQ; and Flavia Krause-Jackson, "Affordable Global Housing Will Cost $11 Trillion," *Bloomberg News*, September 30, 2014.

31. "The Mineral Products Industry at a Glance, 2016 Edition," Mineral Products Association, 2016, 20; and Erwan Garel, Wendy Bonne, and M. B. Collins, "Offshore Sand and Gravel Mining," in *Encyclopedia of Ocean Sciences, 2nd ed.*, John Steele, Steve Thorpe, and Karl Turekian, eds. (New York: Academic Press, 2009), 4162–170.

32. "Dunes and don'ts: the nitty-gritty about sand," *The National*, January 7, 2010.

33. Dobbs, et al., *No Ordinary Disruption*, 18.

34. "Global Trends: Alternative Worlds," US National Intelligence Council, 47.

35. Lee, et al., "Resources Futures," *Chatham House*, xi.

BIBLIOGRAPHY

The list below includes only published books and the most important documents that I used in my research. The many other newspapers, magazines, websites, etc. from which I gleaned information and insight are noted in the Endnotes section.

Abraham, David S. *The Elements of Power: Gadgets, Guns, and the Struggle for a Sustainable Future in the Rare Metal Age.* New Haven: Yale University Press, 2015.

Allman, T. D. *Finding Florida: The True History of the Sunshine State.* New York: Grove Press, 2014.

Asimov, Isaac. *Eyes of the Universe: A History of the Telescope.* Houghton Mifflin Harcourt, 1975.

Banham, Reyner. *A Concrete Atlantis: U.S. Industrial Building and European Modern Architecture.* Cambridge: MIT Press, 1989.

Biddix, David and Chris Hollifield. *Images of America: Spruce Pine.* Mt. Pleasant, SC: Arcadia Publishing, 2009.

Carson, Rachel. *The Edge of the Sea,* reprint ed. New York: Mariner Books, 1998.

Chapman, Emily, et al. "Communities at Risk: Frac Sand Mining in the Upper Midwest," Boston Action Research, September 25, 2014.

Constable, Trevor. "Bagnold's Bluff: The Little-Known Figure Behind Britain's Daring Long Range Desert Patrols," *The Journal of Historical Review* 18, no. 2 (March/April 1999).

Courland, Robert. *Concrete Planet: The Strange and Fascinating Story of the World's Most Common Man-Made Material.* Amherst, NY: Prometheus Books, 2011.

Davenport, Bill, Gerald Voigt, and Peter Deem. "Concrete Legacy: The Past, Present, and Future of the American Concrete Pavement Association," American Concrete Pavement Association, 2014.

Davis, Diana K. *The Arid Lands: History, Power, Knowledge.* Cambridge, MA: MIT Press, 2016.

Dobbs, Richard, James Manyika, and Jonathan Woetzel. *No Ordinary Disruption.* New York: Public Affairs, 2016.

Dolley, Thomas. "Sand and Gravel: Industrial," *US Geological Survey Mineral Commodity Summaries,* January 2016.

Dunn, Richard. *The Telescope.* National Maritime Museum, 2009.

Eisenhower, Dwight D. *At Ease: Stories I Tell to Friends.* Doubleday, 1967.

Floyd, Barbara L. *The Glass City: Toledo and the Industry That Built It.* Ann Arbor: University of Michigan Press, 2014.

Foster, Mark S. *Henry J. Kaiser: Builder in the Modern American West.* Austin: University of Texas Press, 2012.

Freedonia Group. *World Construction Aggregates.* 2016.

————. *World Flat Glass Market Report.* 2016.

Garel, Erwan, Wendy Bonne, and M. B. Collins. "Offshore Sand and Gravel Mining," *Encyclopedia of Ocean Sciences,* 2nd ed., John Steele, Steve Thorpe, and Karl Turekian, eds. New York: Academic Press, 2009.

Gelabert, Pedro A. "Environmental Effects of Sand Extraction Practices in Puerto Rico," papers presented at a UNESCO–University of Puerto Rico workshop entitled "Integrated Framework for the Management of Beach Resources within the Smaller Caribbean Islands," October 21–25, 1996.

Gillis, John R. *The Human Shore: Seacoasts in History,* reprint ed. Chicago: University of Chicago Press, 2015.

————. *The Shores Around Us.* Self-published, 2015.

Global Witness. "Shifting Sand: How Singapore's demand for Cambodian sand threatens ecosystems and undermines good governance," May 2010.

Glover, Alex. "A Brief Review of the History, Geology and Modern Uses of the Minerals Mined in the Spruce Pine Mining District," http://owacc.com/land%20sales/345%20AVERY/345.pdf.

Greenberg, Gary. *A Grain of Sand: Nature's Secret Wonder.* Minneapolis: Voyageur Press, 2008.

Greenberg, Gary, Carol Kiely, Kate Clover. *The Secrets of Sand.* Minneapolis: Voyageur Press, 2015.

Haus, Reiner, Sebastian Prinz, and Christoph Priess. "Assessment of High Purity Quartz Resources," *Quartz: Deposits, Mineralogy and Analytics.* Springer Geology, 2012.

Heiner, Albert P. *Henry J. Kaiser: Western Colossus.* Halo Books, 1991.

International Association of Dredging Companies. *Beyond Sand and Sea.* 2015.

Irving, David. *Hitler's War.* London: Focal Point Publications, 2001.

Kolman, René. "New Land by the Sea: Economically and Socially, Land Reclamation Pays," International Association of Dredging Companies, May 2012.

Kondolf, G. Mathias, et al. "Freshwater Gravel Mining and Dredging Issues," *White Paper Prepared for Washington Department of Fish and Wildlife.* April 4, 2002.

Krane, Jim. *City of Gold: Dubai and the Dream of Capitalism.* New York: St. Martin's Press, 2009.

Krausmann, Fridolin, et al. "Growth in global materials use, GDP and population during the 20th century," *Ecological Economics* 68 (June 10, 2009).

Lee, Bernice, et al. *Resources Futures.* Chatham House, 2012.

Lewis, Tom. *Divided Highways: Building the Interstate Highways, Transforming American Life.* Ithaca, NY: Cornell University Press, 2013.

Macfarlane, Alan and Gerry Martin. *The Glass Bathyscaphe: How Glass Changed the World.* London: Profile Books, 2011.

Maugeri, Leonardo. *Oil: The Next Revolution.* Harvard Kennedy School/Belfer Center for Science and International Affairs, June 2012.

McNeill, Ryan, Deborah J. Nelson, and Duff Wilson. "Water's edge: the crisis of rising sea levels." Reuters, September 4, 2014.

Miodownik, Mark. *Stuff Matters: Exploring the Marvelous Materials That Shape Our Man-Made World.* Boston: Houghton Mifflin Harcourt, 2014.

Morgan, Mike. *Sting of the Scorpion: The Inside Story of the Long Range Desert Group.* Stroud, Glouchestershire: The History Press, 2011.

National Intelligence Council. *Global Trends 2030: Alternative Worlds.* December 2012.

Padmalal, D. and K. Maya. *Sand Mining: Environmental Impacts and Selected Case Studies.* New York: Springer, 2014.

Pearson, Thomas W. *When the Hills Are Gone: Frac Sand Mining and the Struggle for Community.* Minneapolis: University of Minnesota Press, 2017.

Peduzzi, Pascal. *Sand, rarer than one thinks.* United Nations Environment Programme Report, March 2014.

Petroski, Henry. *The Road Taken: The History and Future of America's Infrastructure.* New York: Bloomsbury, 2016.

Pilkey Jr., Orrin H. and J. Andrew G. Cooper. *The Last Beach.* Durham, NC: Duke University Press, 2014.

Ransome, Ernest and Alexis Saurbrey. *Reinforced Concrete Buildings.* New York: McGraw-Hill, 1912.

Ressetar, Tatyana. "The Seaside Resort Towns of Cape May and Atlantic City, New Jersey Development, Class Consciousness, and the Culture of Leisure in the Mid to Late Victorian Era," thesis, University of Central Florida, 2011.

Rundquist, Soren and Bill Walker. "Danger in the Air." Environmental Working Group, September 25, 2014.

Schlanz, John W. "High Pure and Ultra High Pure Quartz," *Industrial Minerals and Rocks*, 7th ed. Society for Mining, Metallurgy, and Exploration, March 5, 2006.

Shixiong Cao et al. "Damage Caused to the Environment by Reforestation Policies in Arid and Semi-Arid Areas of China," *AMBIO: A Journal of the Human Environment* 39, no. 4 (June 2010).

Shurkin, Joel N. *Broken Genius: The Rise and Fall of William Shockley, Creator of the Electronic Age.* New York: Palgrave Macmillan, 2006.

Siever, Raymond. *Sand.* New York: Scientific American Library, 1988.

Skrabec Jr., Quentin. *Michael Owens and the Glass Industry.* Gretna, LA: Pelican, 2006.

BIBLIOGRAPHY

Slaton, Amy E. *Reinforced Concrete and the Modernization of American Building, 1900–1930*. Baltimore, MD: Johns Hopkins University Press, 2001.

Smil, Vaclav. *Making the Modern World: Materials and Dematerialization*. Hoboken, NJ: Wiley, 2013.

Snyder, Laura J. *Eye of the Beholder: Johanees Vermeer, Antoni van Leeuwenhoek, and the Reinvention of Seeing*. New York: W. W. Norton., 2015.

Supreme Court of India. *Deepak Kumar and Others v. State of Haryana and Others*, 2012.

Swift, Earl. *The Big Roads: The Untold Story of the Engineers, Visionaries, and Trailblazers Who Created the American Superhighways*. Boston: Houghton Mifflin Harcourt, 2011.

United Nations Department of Economic and Social Affairs. *World Urbanization Prospects*. 2014.

Weimin Xi, et al. "Challenges to Sustainable Development in China: A Review of Six Large-Scale Forest Restoration and Land Conservation Programs," *Journal of Sustainable Forestry* 33 (2014).

Welland, Michael. *Sand: The Never-Ending Story*. Berkeley: University of California Press, 2009.

Wermiel, Sara. "California Concrete, 1876–1906: Jackson, Percy, and the Beginnings of Reinforced Concrete Construction in the United States," *Proceedings of the Third International Congress on Construction History*. May 2009.

Willett, Jason Christopher. "Sand and Gravel (Construction)," *US Geological Survey Mineral Commodity Summaries*, January 2017.

Wisconsin Department of Natural Resources. *Silica Sand Mining in Wisconsin*. January 2012.

Woodbury, David O. *The Glass Giant of Palomar*. New York: Dodd, Mead, 1970.

Xijun Lai, David Shankman, et al. "Sand mining and increasing Poyang Lake's discharge ability: A reassessment of causes for lake decline in China," *Journal of Hydrology* 519 (2014).

INDEX